Reclaiming the Environmental Debate

Reclaiming the Environmental Debate
The Politics of Health in a Toxic Culture

edited by Richard Hofrichter

The MIT Press
Cambridge, Massachusetts
London, England

Excerpts from *The Corporate Planet* by Joshua Karliner, copyright © 1997 by Joshua Karliner, reprinted with permission of Sierra Club Books; *Toxic Sludge Is Good for You: Lies, Damn Lies, and the Public Relations Industry* by John Stauber and Sheldon Rampton, copyright 1995 by the Center for Media and Democracy, all rights reserved, used by permission of Common Courage Press; *Living Downstream: An Ecologist Looks at Cancer and the Environment* by Sandra Steingraber (extract from chapter 12) copyright 1997 Sandra Steingraber, Ph.D., reprinted by permission of Addison Wesley Longman.

This book was set in Sabon by Achorn Graphic Services, Inc., and was printed and bound in the United States of America.
Printed on recycled paper.

Library of Congress Cataloging-in-Publication Data

Reclaiming the environmental debate : the politics of health in a toxic culture / edited by Richard Hofrichter.
    p.   cm. — (Urban and industrial environments)
  Includes bibliographical references and index.
  ISBN 0-262-58184-5 (hbk. : alk. paper) — ISBN 0-262-58182-5 (pbk. : alk. paper)
  1. Environmental health—Political aspects. 2. Environmental justice. 3. Environmental health—Government policy. 4. Green movement. I. Series. II. Hofrichter, Richard.
RA566.R43   2000
362.1 21—dc21

                                                            99-045525

Contents

Acknowledgments

I would like to thank the following foundations and individuals whose generous support made this book possible: The Jennifer Altman Foundation, The Rockefeller Family Fund, and Carol Bernstein Ferry. Special thanks to John Burt, whose early support and encouragement were essential to the project.

Many people offered their time and skill in discussing or reviewing early drafts of sections of the book. They include Julie Barnet, Joseph Hawes, Grigsby Hubbard, Nancy Meyer, Mary O'Brien, Marilyn Richter, and Sylvia Weber.

Bill Shutkin encouraged me to work with MIT Press and offered many suggestions throughout the process.

To my editor, Clay Morgan, at MIT Press and Enza Vescera, a special thanks, as well as to Robert Gottlieb, series editor, whose critical suggestions significantly improved the book.

I am particularly grateful to the Urban Habitat Program, especially Carl Anthony, for agreeing to house the project, as well as to Viveka for her timely assistance.

Reclaiming the Environmental Debate

1

Introduction: Critical Perspectives on Human Health and the Environment

Richard Hofrichter

We live in a toxic culture that degrades human and environmental health, particularly in communities of the poor and people of color.[1] In workplace, home, communities, and recreational areas, exposure to an expanding array of toxic conditions in the air, water, and soil poses an increasing long-term threat to public health.[2] However, what makes our society a toxic culture are the social arrangements that encourage and excuse the deterioration of the environment and human health.

The traditional understanding of *toxic* is "materials and processes which may cause harm to human health and the environment."[3] My use of the term emphasizes how our way of life and overall social conditions can make communities unhealthy. Thus, elements of toxic culture might include the unquestioned production of hazardous substances, tolerance for economic blight, dangerous technologies, substandard housing, chronic stress, and exploitative working conditions.

We can also think of toxic culture in relation to the consequences of such conditions, recognizing how physical environments and ecosystems are connected to the economic, social, and spiritual health of a community. Toxic culture is also a metaphor for the way language, concepts, rituals, valuation processes, and policies frame the debates over major issues, ignoring the political conflict and relations of power that influence human and community health. These power relations generate their own logic, ideology, myths, symbols, and cultural norms. More important, we are often unaware of their existence.

The acceptance of this culture is manifest in the way many risks to human health are taken for granted, as expected consequences of technological development, or are treated as problems to be managed through public policy. Pesticide-laden and unnutritious foods, acid rain, ozone depletion, air particulates, lead poisoning, water-borne pathogens, chemical spills, and hazardous waste all appear as unfortunate but common occurrences in everyday life.[4] Challenging such a culture will require shifts in public attitudes and awareness so that we can begin to create the social conditions for well-being and build healthy communities.

Need for Revitalized Public Debate

Many people are searching for better ways to create a healthy and sustainable society, and to develop robust public discussion and debate about central concerns in their lives. However, the space for public debate and its quality are diminishing.[5] Experts, economists, and pundits from conservative think tanks dominate the mass media and shape public consciousness about what is politically desirable and possible. The resulting atrophy of imagination often results in narrow and trivial public debate about issues of environmental health. Equally important, many community or oppositional voices have been effectively silenced or excluded from the debate.

Scientists have not been able to evaluate fully and effectively the synergistic, cumulative effects of exposure to multiple toxicants, particularly under conditions of poor health, poverty, and the stresses of other social inequities. While public and scientific recognition of the relationships among deteriorating ecological systems, social conditions, inequality, and patterns of human health has increased over the past century, mainstream portrayals of toxic environments often present their consequences as the inevitable and natural, if unwanted, by-products of a modern industrial society. These are said to be the workings of nature, carelessness, or technological failure. Thus, instead of looking to the roots of ill health in the social arrangements that generate toxicity—poverty, racism, and inequality, for example—policy-makers find themselves, at best, seeking to evaluate each newly identified toxin after it has already arrived on the market, to fund research for cures, or to provide limited and often expensive cleanup programs.

It is not merely the presence of toxic substances per se that harm human health and the environment, thereby creating a toxic culture. Toxic culture arises from the material relations embedded in a social order. Such an order has become dependent on endless development that exploits human and natural resources, including entire ecosystems.[6] Toxic culture turns nature and, increasingly, human health itself into commodities that are no different than other market relations. The use of market discourse in debates over health also diverts attention from how human health is affected and shaped by systems of production and consumption.

Many analysts recognize that improving the health of communities is a thoroughly economic and social enterprise involving housing, transportation, employment, the reduction of poverty, and improvements in the quality of life. Yet, in practice, they rarely make connections among social needs, pollution, health status, and economic systems when exploring the roots of documented patterns of ill health.[7] To do so could lead to questioning the foundations of our market-dominated and increasingly unequal social order of capitalism and expose more clearly the source of health inequities. It is safer to change behavior than to investigate investment decisions, the organization of work, or the way the tax system promotes or degrades the health of communities.

More than a hundred years ago, Frederick Engels demonstrated how the conditions of the working class in England and the distribution of disease were social phenomena associated with politics and power.[8] Evidence indicates that ill health and poor quality of life correlate highly with large differences in the standard of living.[9] Thus, a strong correlation exists between children living in poverty and their susceptibility to environmental hazards and infectious diseases. The issues of production and consumption—how people live in our market-dominated society—also are related to the cultural influences in our daily life.[10] Social theorist John Tomlinson explains the cultural dimension of domination this way:

To grasp the order of "blame" in cultural domination we have to think of capitalism as something wider than the practices of individual capitalist organizations. . . . This wider view . . . represents it as something within which the routine practices both of ordinary people and of individual capitalist organizations are locked. It is in this wide sense that we can speak of a "culture" of capitalism.

. . . It is not individual practices we are blaming, but a contextualizing structure: capitalism not just as economic practices but as the *central (dominant) positioning of economic practices* [emphasis in original] within the social ordering of collective existence.[11]

The dichotomy between threats to human health and our understanding of actions to prevent these threats calls for an explanation of the context in which social decisions and individual choices occur. How, for example, can we avoid making the critical analysis linking social inequalities, health, and environmental degradation? What inhibits a more coordinated attack on the roots of ill health in social conditions? How can those who hold opposing perspectives effectively articulate their views?

Mainstream discourses on environmental and human health typically ignore the need for a major structural transformation to prevent increasing levels of toxicity, ill health, and inequitable living conditions. Yet these discourses shape consciousness and affect the way people describe, interpret, represent, experience, construct, and ultimately legitimate social reality. They promote or justify forms of economic and social development (including the production, use, and disposal of toxic substances) that create ill health.

Western concepts about progress, development, private property, economic rationality, risk, science, and individualism in everyday life inhibit critical attention to the social determinants of health. These concepts contribute to the emphasis on consumption over production and private over public gain, thereby transforming collective issues into personal ones. More important, they remain deeply ingrained in institutionalized practices that constrain society's willingness to rethink what is desirable and possible.

The potential for successful resistance and disruption of the discourse of capital is always present. The mainstream talk about progress, markets, consumption, and technological solutions is becoming increasingly irrelevant and disconnected from people's experience with environmental and health concerns. Opportunities for resistance grow as consent to the perspective of capital becomes more fragile under conditions in which social needs remain unmet and corporate images of innocence jar with reality.[12] The moment at which such disruption may occur is unpredictable and volatile, as the seductiveness of capitalist culture exhausts itself in meaningless repetition and incoherence. At any historical moment, social forces can erupt that shift the political terrain, because corporate hegemony is always in a state of change, ever needing to be recreated.

Purpose of This Volume

Reclaiming the Environmental Debate is about the contradictions within, and resistance to, some of the central debates and practices about human health and the environment as they develop in different contexts such as urban redevelopment, siting of industrial facilities, advertising, uses of space, risk assessment, resource extraction, disposal of toxic waste, and occupational health and safety. A central theme is that effective challenges to toxic culture, as well as the potential for creating a compelling vision of a healthy society, grounded in everyday work and life, require reframing objectives so as to produce broad, comprehensive social change. Thus, instead of emphasizing policy reform, medical advances, and individual behavior, society would focus on the causes of ill health associated with the production, use, and disposal of resources and, more important, social inequality.

The contributors to this volume reflect diverse voices and critical perspectives. Their essays explore contemporary interpretations of threats to human health and the environment and the responses to them. The essays range from critiques of influential discourses and practices to the development of strategies for creating healthy communities. The connection among the essays is their broad challenge to the core ideas that support a toxic culture. They rethink the definition of a healthy society and offer creative approaches to organizing against toxic culture.

In seeking contributors, I discovered that there were many paths, levels, and approaches linking theory and practice, most of them compatible. Some examine the broad conceptual, historical context that would guide progressive action. Others expose patterns of corporate behavior and influence peddling that suggest caution in evaluating scientific research on issues of environmental health. Community activists, in daily battles with corporations and government agencies, tell a vastly different story about the politics of health in their communities than that which appears in the mainstream mass media, particularly on the effects of the continuing poisoning of their communities and explanations for them.

The contributors also represent many backgrounds: grassroots organizing, public health, ecology, cultural studies, environmental politics, and muckraking journalism, to name a few. Within their wide range of views (some disagreeing with my own), they exemplify provocative recent thinking about progressive possibilities for rebuilding healthy, clean communities. They seek to reframe traditional debates, questioning understandings about risk, space, economics, culture, and science that are taken for granted. They examine such questions as why and how social constructions about environmental deterioration and ill health become interpreted as inevitable, natural, and organic, thereby restricting the range of social choices. What are people doing that might provide direction for a more democratic process in building and sustaining a healthy society? What is necessary to build public support? And what accounts for the discrepancy between knowledge and action in improving human health?

The common threads that tie the articles together are their recognition of the political and cultural dynamics that influence the health of communities. In different ways, they all attempt to overcome the influence that powerful interests have in distorting the sources of ill health and the responses to it. They promote and legitimate organized activity in planning for healthy communities and responding collectively to toxic culture.

The essays are also distinctive in that each assumes a broad definition of the term "toxic" and the conditions required for a healthy community. In addition, because *toxic culture* refers partly to the language and codes of understanding we use in discussing the health and illness of communities, the authors in many instances avoid conventional policy debates and consider instead the nature of communication and protest in regard to issues of human and environmental health.

Thus this volume is not primarily focused on policy alternatives, competing truth claims, or purely fact-based accounts. Nor do any of the articles demonstrate a direct correlation between a specific form of resistance and an outcome. The book does not seek to provide "solutions." Instead, the essays seek to capture the ideological nuances of how society thinks about and portrays the issues of a toxic culture. The contributors find value in raising questions that are not easily answered and in describing forms of opposition that emphasize changes in consciousness or paradigm shifts rather than programmatic change.

My purpose is to stimulate thinking by having the reader examine the idea of toxic culture through viewpoints that are not often encountered— whether by presenting the direct experiences of community activists or critical perspectives on commonplace talk in the media about economics and risk. This book has an interdisciplinary perspective that interweaves the personal, the political, and the academic.

Organization of the Book

The book is divided into three parts. Briefly, the first part identifies the institutions and philosophies that dominate the debate on human health and the environment. The essays reveal the flawed assumptions, premises, and methodologies that support the prevailing mindset and identify the consequences. Essays in the second part consider how specific types of language and imagery that corporations and technocrats use distort the issues and promote limited policy and individualistic rather than collective responses to the ill health of communities. The third part presents case studies of organized resistance to living in a toxic culture, primarily at the grassroots level.

Part I is about contesting the concepts, theories, and language that govern debate and affect the shaping of consciousness about human health in toxic environments. The contributors consider the perspectives that influence how people interpret environmental and public health concerns, judge the basis on which decisions are made, and establish visions of a healthy society. Their essays show how the contradictions within major interpretive frameworks (about progress, endless growth, place, production, consumption, and risk) provide opportunities for progressive social transformation.

Specifically, what are the implications of failing to challenge dominant perspectives? The absence of a broader challenge to a system that generates ill health typically channels public attention toward a limited legal and policy terrain on which to engage the issues. For example, attributing increasing rates of cancer to individual behavior or biological processes instead of social conditions will most likely lead to managerial or moral responses more than political ones. This forestalls organized action to combat the true sources in social and economic structures and power arrangements that drive the capitalist social order.

Proposed solutions for communities threatened by hazardous substances, for example, stress the application of science, medicine, and technology, not fundamental change to address the social determinants of health such as poverty, unemployment, and poor housing conditions. Instead of demanding clean water, which might expand the scope of public health intervention, people buy filters for their sinks or drink bottled water. Instead of challenging the direction of social technologies, medical professionals suggest changes in consumption habits.

Policy-makers and corporate interests seek to manage, regulate, and treat ill health through control of pollution. Questions about the deep-rooted causes of the pollution receive little attention. Debates about the nature of, sources of, and solutions to an environmental health problem or crisis always involve political struggle and social interpretation, although they often appear technical. Even use of the term "environmental problem" is already a capitulation. It ignores an underlying historical narrative that links health and ecology to the social order that defines well being in contemporary capitalist culture. The idea of an isolated "environmental problem" narrows the scope and method of public health intervention. What is needed is a more comprehensive view that considers the health of entire communities.

Such a view requires challenging those ideologies that limit critical thinking. While issues of toxic waste, global climate change, nuclear power, and ozone depletion are of great public concern, they often become entrapped in policy discussions whose subtext is shareholder interests. Interventions in these debates may offer an opportunity for articulating a progressive vision that accentuates the recovery of public space for political action. Involvement in these issues may also help community residents become more knowledgeable about the requirements for a democratic society, strengthening their capacities to make social choices in place of experts.

One of the central objectives in reconstructing a vision of a healthy society is to recover (borrowing a term from Frederic Jameson) the political unconscious,[13] uncovering the repressed origins of illness in capitalist social relations by tracing the history of the conditions under which people live, including the production of hazardous substances that produce ill health. One important task in this regard is to demonstrate the relationship between a healthy society and social justice and equity.

Achieving that task also requires that we examine the way corporate imagery, language, and practice, as well as the language of technoscience influence public debate on health and environmental crises. The essays in Part II explore the role of advertising, media, and public relations in this debate. They identify the phenomenon called "corporate greenwashing" and show how the subtle integration of corporate-speak with technocratic, scientific practice affects the ability of the public to play a role in protecting and improving health and the environment.

Global corporations are shaping public debate more aggressively than ever before on issues of ecology and environmental health. Their unprecedented influence over the mass media enhances their capacity to appropriate "green" symbols/images, direct the use of university research funds, and often co-opt national environmental organizations.[14] Consequently, ecological images and narratives presented by mainstream environmentalists are mostly empty of social and historical context, disconnected from broader social justice perspectives, and easily co-opted.

Capital requires, as a matter of sustaining growth, a homogenized, universal consumer culture—a world where the logic of buying and selling and the exploitation of people and natural resources insinuates itself into every area of public and private life.[15] Global corporations, as manifestations of capital, influence the character of the public, social space in which people interpret and give meaning to the world through language, symbols, imagery, and cultural practices. As David Korten argues,

The architects of the corporate global vision seek a world in which universalized symbols created and owned by the world's most powerful corporations replace the distinctive cultural symbols that link people to particular places, values, and human communities. Our cultural symbols provide an important source of identity and meaning. . . . They arouse . . . [a] sense of responsibility for the health and well-being of our community and its distinctive ecosystem. When control of our cultural symbols passes to corporations, we're essentially yielding to them the power to define who we are.[16]

Korten notes that "television has already been wholly colonized by corporate interests. . . . The goal is not simply to sell products . . . [but]

to create a political culture that equates the corporate interest with the human interest in the public mind."[17] However, the exercise of capital's power is not always observable or explicit; it operates in routine practices within the institutions of everyday life—the design of buildings, the layout of streets, or the direction of technological developments—not merely economic production or the mass media.[18]

Even while voicing growing concern about ecological degradation, corporations conceal their role in this devastation and seek to neutralize opposition by implying that responsible, sympathetic action is a priority. They present to the public an idea of progress defined mainly in technological and economic terms, such as contract, free markets, free exchange, and individualism, excluding ideas of justice or cooperation. To the extent that people fail to identify themselves primarily as active citizens who can imaginatively create a future society, this corporate-speak fosters their identity as unorganized consumers who function in the marketplace.

Rarely do the symbols and language of the mainstream environmental movement subvert the social order in a way that questions the social sources of environmental degradation and its inequitable effects on oppressed populations. The stories and histories created by capitalist institutions conflict with those that arise out of people's history and daily life because the former exist primarily as a means to create markets and market relationships. Corporate stories remain powerful to the extent that they are ubiquitous, unrecognized, or otherwise taken for granted. It requires more than the exposure of lies to counter the discourses that permeate the culture. Consider, for example, how gross national product and stock prices often serve as indicators of national success.

Corporations determine how they will represent themselves to themselves and to others in order to achieve their economic and political objectives, which include expanding markets, controlling capital and technology, socializing costs and risks so they are borne by society, sustaining unfettered property and investment rights, creating new forms of property, and subordinating labor. (Some of the worst polluters in the United States use ecological images in their ads without losing credibility.) Corporations need not always justify themselves explicitly; their economic power through control of capital itself gives their ideology a certain legitimacy. However, as Australian social critic Arran Gare notes, they do need to "[devalue] that which does not serve as an instrument of the international economy"[19] and create identities through consumption rather than the labor process.

Capital, with its trade associations, public relations firms, access to the media, front groups, and powerful lobbyists, is much better organized than most citizens. Increasingly, universities and university scientists are on the payrolls of large corporations. In public schools across the country, corporations are increasingly insinuating themselves into curricula, presenting an innocent face of benign activity on behalf of humanity. However, there are limits and contradictions. Many community organizations are developing versions of history that counter the institutionalized accounts of transnational capital, versions derived from radically different experience.

There are today innovative examples of resistance and community activism, primarily in places where people are most affected by exposure to toxic conditions. Some of these are described in part III. A growing grassroots activism is questioning corporate interpretations of environmental health issues that block the narratives of those campaigning for a more just social order. Such activism expands the boundaries of legitimate political action.[20] Communities have less tolerance than ever before for contamination in their midst. Environmental justice organizations have been coalescing for years to battle corporations directly in court and in the streets, challenging prevailing cultural symbols and belief systems.

New coalitions are beginning to subvert corporate imagery and meaning and are challenging corporate representations, myths, illusions, and broken promises. Many organizations and individuals, as they have done in the past, are finding innovative ways to tell their stories to one another and to the public at large. In the past their activities were excluded from mainstream media and other systems of representation. These stories of people who value healthy communities and workplaces can become a collective, shared narrative about development, the future, and social wellbeing. Reframed narratives can also assist critical rethinking of central tenets of the social order and help groups develop the collective strategies to challenge them. As communications theorist Stanley Deetz, quoting A. Simonds, argues, "Political competence in modern society 'means not just access to information but access to the entire range of skills required to decode, encode, interpret, reflect upon, appraise, contextualize, integrate, and arrive at decisions respecting that information.' "[21] At stake in these struggles is nothing less than human health and the survival of the ecosystems that sustain all life.

Some contributors to part III in this volume describe the development of cultural projects through collaborations between artmakers and grassroots organizations that communicate their own stories and vision of healthy communities and economic and environmental justice through innovative theater, performance, videography, and other cultural endeavors.

Contemporary cultural activist strategies dramatize injustices in such a way that they can be recognized as public, collective issues and experienced as shared, rather than isolated, problems. Such strategies acknowledge and reinforce the voices of those who experience these injustices, emphasizing that the voices in some sense represent the shared experience of all people living in capitalist societies.

To bypass mass-media filtering, co-optation, and spectacle, people in many parts of the United States build political unity and control their images by communicating their concerns directly, without media filtering and framing. To revitalize social imagination, many progressive organizations produce new social theories for envisioning a reconstructed society, affirming unofficial and more accurate versions of history and initiating a different path toward human and social development.[22] They have the potential to ignite critical consciousness, build solidarity and political awareness, and interrogate or disrupt dominant systems of power. This is especially true when these organizations express community identity, visions of the future, and aspirations through the community's own symbols, language, history, and stories, in a socially conscious, uncompromising, public way, rooted in social life.[23,24]

Culture is formed through a living, breathing, historical, social process in which everyone plays a role. It represents the way a community of people views itself. Politics always serves as a theme, a narrative in people's lives, and almost all political activity is a cultural expression. The critical perspectives expressed in this volume offer a starting point for creating a dialogue that calls into question the toxic culture generated by corporate capital.

Notes

1. Robert Bullard, *Unequal Protection* (San Francisco: Sierra Club Books, 1994); Benjamin Goldman, *The Truth About Where You Live: An Atlas for Action on Toxins and Mortality* (New York: Random House, 1992). See also Gerald Torres, "Environmental Burdens and Democratic Justice," *Fordham Urban Law Journal* 21(3) (Spring 1994):431–460; Richard J. Wilkinson, *Unhealthy Societies: The Affliction of Inequality* (New York: Routledge, 1996); Gerald Markowitz and David Rosner, "Pollute the Poor," *The Nation* (July 6, 1998):8; Rachel's Environment and Health Weekly, "Philadelphia Dumps on the Poor," no. 595 (April 23, 1998).

2. Eric Chivian et al., *Critical Condition: Human Health and the Environment* (Cambridge, Mass.: MIT Press, 1994); Gar Alperovitz et al., *Index of Environmental Trends: An Assessment of Twenty-One Key Environmental Indicators in Nine Industrialized Countries over the Past Two Decades* (Washington, D.C.: National Center for Economic Alternatives, 1995); David Pimentel, "Ecology of Increasing Disease," *BioScience* 48(1) (October, 1998):817; Eric Mann, *L.A.'s Lethal Air* (Los Angeles: Labor/Community Strategy Center, 1991); Gary Cohen and John O'Connor, *Fighting Toxics* (Washington, D.C.: Island Press, 1990).

3. Robert Gottlieb, *Reducing Toxics: A New Approach to Policy Making and Industrial Decisionmaking* (Washington, D.C.: Island Press, 1995), p. 1.

4. For a detailed discussion of the threats posed on these issues, see J. L. Pirkle et al., "The Decline of Blood Lead Levels in the United States: The National Health and Nutrition Examination Surveys," *Journal of the American Medical Association,* 272 (1994):284–291; D.V. Bates, "The Effects of Air Pollution on Children." *Environmental Health Perspectives* 103(6) 1995:49–53; Centers for Disease Control and Prevention, "Surveillance for Asthma-United States, 1960–95," *Morbidity and Mortality Weekly Report* 47, SS-1, (1998) p. 1; Leonard Legault et al., Ninth Biennial Report on Great Lakes Water Quality (Washington, D.C.: International Joint Commission, 1998).

5. See, for example, Herbert I. Schiller, *Culture Inc.: The Corporate Takeover of Public Expression* (New York: Oxford Univ. Press, 1989); Robert W. McChesney, Ellen Meiksins Wood, and John Bellamy Foster (eds.), *Capitalism and the Information Age: The Political Economy of the Global Communication Revolution* (New York: Monthly Review Press, 1998); and Barry Sanders, *The Private Death of Public Discourse* (Boston: Beacon Press, 1998).

6. Martin O'Connor, "On the Misadventures of Capitalist Nature," in Martin O'Connor (ed.), *Is Capitalism Sustainable?: Political Economy and the Politics of Ecology* (New York: Guilford Press, 1994), pp. 125–151; Vincente Navarro, *Crisis, Health, and Medicine: A Social Critique* (New York: Tavistock Publications, 1986); Pratap Chatterjee and Matthias Finger, *The Earth Brokers: Power, Politics, and World Development* (New York: Routledge, 1994).

7. See Meredeth Turshen, *The Politics of Public Health* (Rutgers, NJ: Rutgers Univ. Press, 1989).

8. Frederick Engels, *The Condition of the Working Class in England* (Stanford Univ. Press (1958). (Reprint of the 1848 edition.)

9. See Wilkinson, *Unhealthy Societies;* Alan R. Tarlov, "Social Determinants of Health: The Sociobiological Translation," in *Health and Social Organization: Towards a Health Policy for the 21st Century,* David Blane, Eric Brunner and Richard Wilkinson, eds. (New York: Routledge, 1996), pp. 71–93; Eileen Stillwagon, *Stunted Lives, Stagnant Economies: Poverty, Disease, and Under-development* (New Brunswick, NJ: Rutgers University Press, 1998).

10. See Michael E. Brown, *The Production of Society* (Totowa, NJ: Rowman and Littlefield, 1986); Henri Lefebvre, *Everyday Life in the Modern World* (New York: Harper and Row, 1971).

11. John Tomlinsin, *Cultural Imperialism* (Baltimore, Md: Johns Hopkins Univ. Press, 1991), p. 168.

12. Sharon Beder, *Global Spin: The Corporate Assault on Environmentalism* (White River Junction, VT: Green Books & Chelsea Green Publishing, 1998). See also Timothy Luke, *Ecocritique: Contesting the Politics of Nature, Economy, and Culture* (Minneapolis: University of Minnesota Press, 1997; Robert Goldman and Stephen Papson, *Sign Wars: The Cluttered Landscape of Advertising* (New York: The Guilford Press, 1996).

13. Frederic Jameson, *The Political Unconscious: Narrative as a Socially Symbolic Act* (Ithaca, NY: Cornell University Press, 1981).

14. See Beder, *Global Spin.*

15. Frederic Jameson, *Postmodernism or The Cultural Logic of Late Capitalism* (Durham, NC: Duke University Press, 1991); Richard J. Barnet and John Cavanagh, *Global Dreams: Imperial Corporations and the New World Order* (New York: Simon & Schuster, 1994).

16. David Korten, *When Corporations Rule the World* (West Hartford, CT: Kumarian Press, 1995), p. 158.

17. Korten, p. 149. For a detailed analysis of corporate relations activities and human health see Judith Richter, "Engineering of Consent: Uncovering Corporate PR" (Dorset, UK: The Corner House, Briefing 6, March, 1998).

18. Antonio Gramsci, *Selections From The Prison Notebooks* (New York: International Publishers, 1971); Raymond Williams, *Marxism and Literature* (London: Oxford University Press, 1977).

19. Arran E. Gare, *Postmodernism and the Environmental Crisis* (New York: Routledge, 1995), p. 11.

20. See Daniel Faber, ed., *The Struggle for Ecological Democracy: Environmental Justice Movements in the United States* (New York: The Guilford Press, 1998).

21. Stanley Deetz, *Democracy in the Age of Corporate Colonization: Developments in Communication and the Politics of Everyday Life* (Albany, NY: State University of New York Press, 1992), p. 2.

22. Andrew Szasz, *Ecopopulism: Toxic Waste and the Movement for Environmental Justice* (Minneapolis: University of Minnesota Press, 1994).

23. See, for example, Bron Raymond Taylor (ed.), *Ecological Resistance Movements: The Global Emergence of Radical and Popular Environmentalism* (Albany, N.Y.: State Univ. of New York Press, 1995); David Camacho, *Environmental Injustices, Political Struggles: Race, Class and the Environment* (Durham, NC: Duke University Press, 1998); Faber, *Struggle for Ecological Democracy*.

24. The civil rights movement, the struggles for women's rights, and the fight against AIDS are strong examples in which cultural expression played a crucial role, particularly music and street theater.

I

Challenging Current Perspectives

2

The Social Production of Cancer: A Walk Upstream

Sandra Steingraber

In 1979, in between my sophomore and junior year in college, I was diagnosed with bladder cancer. Four years later, while a doctoral student in biology, I took the train from Ann Arbor, Michigan, to central Illinois for an appointment with my original urologist. This particular appointment was destined to turn out fine: there were no recurrences. What I remember most clearly is my journey there by train.

Something about the landscape changes abruptly between northern and central Illinois. I am not sure what it is exactly, but it happens right around the little towns of Wilmington and Dwight. The horizon recedes, and the sky becomes larger. Distances increase, as though all objects are slowly moving away from each other. Lines become more sharply drawn. These changes always make me restless and when driving, I drive faster. But since I am in a train, I close the book I am reading and begin impatiently straightening the pages of a newspaper strewn over the adjacent seat.

That is when my eye catches the headline of a back-page article: Scientists Identify Gene Responsible for Human Bladder Cancer. Pulling the newspaper onto my lap, I stare out the window and become very still. It is only early evening, but the fields are already dark, a patchwork of lights quilted over and across them. They have always soothed me. I look for signs of snow. There are none. Finally, I read the article.

Researchers at the Massachusetts Institute of Technology, it seems, had extracted DNA from the cells of a human bladder tumor and used it to transform normal mouse cells into cancerous ones.[1] Through this process, they located the segment of DNA responsible for the transformation. By comparing this segment with its unmutated form in noncancerous human cells, they were able to pinpoint the exact alteration that had caused a respectable gene to go bad.

In this case, the mutation turned out to be a substitution of one unit of genetic material for another in a single rung of the DNA ladder. Namely, at some point during DNA replication, a double-ringed base called guanine was swapped for the single-ringed thymine. Like a typographical error in which one letter replaces another—*snow* instead of *show, block* instead of *black*—the message sent out by this gene was utterly changed. Instead of instructing the cell to manufacture the amino acid glycine, the altered gene now specified valine. (Nine years later, other researchers would determine that this substitution alters the structure of proteins involved in signal transduction—the crucial line of communication between the cell membrane and the nucleus that helps coordinate cell division.)

Guanine instead of thymine. Valine instead of glycine. I look away again—this time at my face superimposed over the landscape by the window's mirror. If, in fact, this mutation was involved in my cancer, when did it happen? Where was I? Why had it escaped repair? I had been betrayed. But by what?

Thirteen years later, I possess a bulging file of scientific articles documenting an array of genetic changes involved in bladder cancer.[2] Besides the oncogene just described, two tumor suppressor genes, p15 and p16, have also been discovered to play a role. Their deletion is a common event in transitional cell carcinoma, the kind of cancer I had. Mutations of the famous p53 tumor suppressor gene, with guest-star appearances in so many different cancers, have been detected in more than half of invasive bladder tumors. Also associated with transitional cell carcinomas are surplus numbers of growth factor receptors. Their overexpression has been linked to the kinds of gross genetic injuries that appear near the end of the malignant process.

The nature of the transaction between these various genes and certain bladder carcinogens has likewise been worked out in the years since a newspaper article introduced me to the then-new concept of oncogenes. Consider, for example, that redoubtable class of bladder carcinogens called "aromatic amines"—present as contaminants in cigarette smoke;

added to rubber during vulcanization; formulated as dyes for cloth, leather, and paper; used in printing and color photography; and featured in the manufacture of certain pharmaceuticals and pesticides.[3] Aniline, benzidine, naphthylamine, and o-toluidine are all members of this group. The first reports of excessive bladder cancers among workers in the aniline dye industry were published in 1895. More than a century later, we now know that anilines and other aromatic amines ply their wickedness by forming DNA adducts in the cells of the tissues lining the bladder, where they arrive as contaminants of urine.

We also now know that aromatic amines are gradually detoxified by the body through a process called "acetylation." Like all such processes, it is carried out by a special group of detoxifying enzymes whose actions are controlled and modified by a number of genes. People who are slow acetylators have low levels of these enzymes and are at greater risk of bladder cancer from exposure to aromatic amines. Members of this population can be readily identified because they bear significantly higher burdens of adducts than fast acetylators at the same exposure levels.[4] These genetically susceptible individuals hardly constitute a tiny minority: more than half of Americans and Europeans are estimated to be slow acetylators.

Very likely, I am one. You may be one, too.

We know a lot about bladder cancer.[5] Bladder carcinogens were among the earliest human carcinogens ever identified, and one of the first human oncogenes ever decoded was isolated from some unlucky fellow's bladder tumor. Sadly, all of our knowledge about genetic mutations, inherited risk factors, and enzymatic mechanisms has not been translated into an effective campaign to prevent the disease. The fact remains that the overall incidence rate of bladder cancer increased 10 percent between 1973 and 1991. Increases are especially dramatic among African Americans: among black men, bladder cancer incidence has risen 28 percent since 1973, and among black women, 34 percent. Somewhat less than half of all bladder cancers among men and one-third of all cases among women are thought to be attributable to cigarette smoking, which is the single largest known risk factor for this disease.[6] The question thus still remains: What is causing bladder cancer in the rest of us, the majority of bladder cancer patients, for whom tobacco is not a factor?

I also possess another bulging file of scientific articles. These concern the continuing presence of known and suspected bladder carcinogens in rivers, groundwater, dump sites, and indoor air. For example, industries reporting to the Toxics Release Inventory disclosed environmental releases of the aromatic amine o-toluidine that totaled 14,625 pounds in 1992 alone.[7] Detected also in effluent from refineries and other manufacturing plants, o-toluidine exists as residues in the dyes of commercial textiles, which may, according to the *Seventh Annual Report on Carcinogens,* published by the U.S. Department of Health and Human Services, expose members of the general public who are consumers of these goods: "The presence of o-toluidine, even as a trace contaminant, would be a cause for concern."[8] A 1996 study investigated a sixfold excess of bladder cancer among workers exposed years earlier to o-toluidine and aniline in the rubber chemicals department of a manufacturing plant in upstate New York.[9] Levels of these contaminants are now well within their legal workplace limits, and yet blood and urine collected from current employees were found to contain substantial numbers of DNA adducts and detectable levels of o-toluidine and aniline. Another recent investigation revealed an eightfold excess of bladder cancer among workers employed in a Connecticut pharmaceuticals plant that manufactured a variety of aromatic amines.[10]

What my various file folders do not contain is a considered evaluation of all known and suspected bladder carcinogens—their sources, their possible interactions with each other, and our various routes of exposure to them. Trihalomethanes—common contaminants of chlorinated tap-water—have been linked to bladder cancer, as has the dry-cleaning solvent and sometime-contaminant of drinking-water pipes, tetrachloroethylene. I possess individual reports on each of these topics. What I do not have is a comprehensive description of how all these substances behave in combination. What are the risks of multiple trace exposures? What happens when we drink trihalomethanes, absorb aromatic amines, and inhale tetrachloroethylene? Furthermore, what is the ecological fate of these substances once they are released into the environment? What happens when dyed cloth, colored paper, and leather goods are laundered, landfilled, or incinerated? And why, almost a century after some of them were so identified, do powerful bladder carcinogens such as amine dyes continue to be manufactured, imported, used, and released into the environment? However improved the record of effort to regulate them, why have they all not been replaced by safer substitutes? To my knowledge, these questions remain largely unaddressed by the cancer research community.

Biased Focus

Genes

Several obstacles, I believe, prevent us from addressing cancer's environmental roots. An obsession with genes and heredity is one.

Cancer research currently directs considerable attention to the study of inherited cancers.[11] Most immediately, this approach facilitates the development of genetic testing, which attempts to predict an individual's risk of succumbing to cancer, based on the presence or absence of certain genetic alterations.

Hereditary cancers, however, are the rare exception. Collectively, fewer than 10 percent of all malignancies are thought to involve inherited mutations.[12] Between 1 and 5 percent of colon cancers, for example, are of the hereditary variety, and only about 15 percent exhibit any sort of familial component.[13] The remaining 85 percent of colon cancers are officially classified as "sporadic," which essentially means that we don't know what causes them.[14] Breast cancer also shows little connection to heredity (probably between 5 and 10 percent).[15] Finding "cancer genes" is not going to prevent the great majority of cancers that develop.

Moreover, even when inherited mutations do play a role in the development of a particular cancer, environmental influences are inescapably involved as well. Genetic risks are not exclusive of environmental risks. Indeed, the direct consequence of some of these damaging mutations is that people become even more sensitive to environmental carcinogens. In the case of hereditary colon cancer, for example, what is passed down the generations is a faulty DNA repair gene.[16] Its human heirs are thereby rendered less capable of coping with environmental assaults on their genes or repairing the spontaneous mistakes that occur during normal cell division. These individuals thus become more likely to accumulate the series of acquired mutations needed for the formation of a colon tumor.

Cancer incidence rates are not rising because we are suddenly sprouting new cancer genes. Rare, heritable genes that predispose their hosts to cancer by creating special susceptibilities to the effects of carcinogens have undoubtedly been with us for a long time. The ill effects of some of these genes might well be diminished by lowering the burden of environmental carcinogens to which we are all exposed. In a world free of aromatic amines, for example, being born a slow acetylator would be a trivial issue, not a matter of grave consequence. The inheritance of a defective carcinogen-detoxifying gene would matter less in a culture that did not tolerate carcinogens in air, food, and water. By contrast, we cannot change our ancestors. Shining the spotlight on inheritance focuses us on the one piece of the puzzle we can do absolutely nothing about.

Lifestyle

Risks of lifestyle are also not independent of environmental risks. Yet, public education campaigns about cancer consistently accent the former and ignore the latter. I collect the colorful pamphlets on cancer that are made available in hospitals, clinics, and waiting rooms. When I was teaching introductory biology and also spending many hours in doctors' offices, I began to compare the descriptions of cancer in the tracts displayed in the racks above the magazines with the chapter on cancer provided in my students' textbook. Here are some of my findings.

On the topic of how many people get cancer, a pink and blue brochure published by the U.S. Department of Health and Human Services offers the following:

Good News: Everyone does not get cancer. 2 out of 3 Americans never will get it.[17]

Whereas, according to *Human Genetics: A Modern Synthesis:*

One of three Americans will develop some form of cancer in his or her lifetime, and one in five will die from it.[18]

(Since these materials were published, the proportion of Americans contracting cancer has risen from 30 to 40 percent.)

On the topic of what causes cancer, the brochure states:

In the past few years, scientists have identified many causes of cancer. Today it is known that about 80% of cancer cases are tied to the way people live their lives.

Whereas the textbook contends:

As much as 90 percent of all forms of cancer is attributable to specific environmental factors.

In regard to prevention, the brochure emphasizes individual choice and responsibility:

You can control many of the factors that cause cancer. This means you can help protect yourself from the possibility of getting cancer. You can decide how you're going to live your life—which habits you will keep and which ones you will change.

The genetics book presents a somewhat different vision:

Because exposure to these environmental factors can, in principle, be controlled, most cancers could be prevented. . . . Reducing or eliminating exposures to environmental carcinogens would dramatically reduce the prevalence of cancer in the United States.

The textbook identifies some of these carcinogens, the routes of exposure, and the types of cancer that result. In contrast, the brochure emphasizes the importance of personal habits, such as sunbathing, that raise one's risk of contracting cancer. Thus, in my students' textbook, vinyl chloride is identified as a carcinogen to which workers making polyvinyl chloride (PVC) are exposed, whereas in the brochure, occupations that involve working with certain chemicals are called a risk factor. The textbook declares that "radiation is a carcinogen." The brochure advises us to "avoid unnecessary X-rays." Both emphasize the role of diet and tobacco.

In its ardent focus on lifestyle, the Good News brochure is typical of the educational pamphlets in my collection. By emphasizing personal habits rather than carcinogens, they present the cause of the disease as a problem of *behavior* rather than one of *exposure* to disease-causing agents. At its best, this perspective can offer us practical guidance and the reassurance that there are actions we as individuals can take to protect ourselves. (Not smoking, rightfully so, tops this list.) At its worst, the lifestyle approach to cancer is dismissive of hazards that lie beyond personal choice. A narrow focus on lifestyle—like a narrow focus on genetic mechanisms—obscures cancer's environmental roots. It presumes that the continuing contamination of our air, food, and water is an immutable fact of the human condition to which we must accommodate ourselves. When we are urged to "avoid carcinogens in the environment and workplace," this advice begs the question. Why must there be known carcinogens in our environment and at our job sites?

Cancer is certainly not the first disease to inspire this kind of message. In 1832, at the height of an epidemic, the New York City medical council announced that cholera's usual victims were those who were imprudent, intemperate, or prone to injury by the consumption of improper medicines.[19] Lists of cholera prevention tips were posted publicly. Their advice ranged from avoiding drafts and raw vegetables to abstaining from alcohol. Maintaining "regular" habits was also said to be protective. Decades later, improvements in public sanitation would bring cholera under control, and the pathogen responsible for the disease would finally be isolated by the bacteriologist Robert Koch in 1883. Of course, the behavioral changes urged by the 1832 handbills were not all without merit: uncooked produce, as it turned out, was an important route of exposure, but it was a fecal-borne bacteria—and not a salad-eating lifestyle—that was the cause.

The orthodoxy of lifestyle today finds its full expression in the public educational literature on breast cancer. Scores of cheerful pamphlets exhort women to exercise, lower the fat in their diets, perform breast self-examinations, ponder their family history, and receive regular mammograms. "Delayed childbirth" (after age twenty) is frequently mentioned as a risk factor. (I have never seen "prompt childbirth" in the accompanying list of cancer prevention tips—undoubtedly because such advice would be tantamount to advocating teenage pregnancy.)[20]

By itself, a lifestyle approach to preventing breast cancer is inadequate.[21] First, the majority of breast cancers cannot be explained by lifestyle factors, including reproductive history. We need to look elsewhere for the causes of these cancers. Second, mammography and breast self-examinations are tools of cancer detection, not acts of prevention. The popular refrain "Early detection is your best prevention!" is a non sequitur: Detecting cancer, no matter how early, negates the possibility of preventing cancer. At best, early detection may make cancer less fatal.

Finally, the adage that high-fat Western diets are the cause of breast cancer has not yet been supported by data.[22] Dietary fat has long been a centerpiece of study in the investigation of breast cancer risk. Yet, several long-term studies have indicated that dietary fat is unlikely to play a major role by itself.[23] Rather than continuing to focus single-mindedly on the absolute quantity of fat consumed, several researchers have called for a more refined, ecological approach to diet.[24] Two obvious starting points would be to assess the link between breast cancer and diets high in animal fat and to launch a definitive investigation into the extent to which various kinds of fats are contaminated by carcinogens. We already know with certainty that animal-based foods are our main route of exposure to organochlorine pesticides and dioxins.[25]

Even reproductive choices have environmental implications. Breasts, for example, do not complete their development until the last months of a woman's first full-term pregnancy. During this time, the latticework of mammary ducts and lobules differentiates into fully functioning secretory cells. This process of specialization permanently slows the rate of mitosis, dampens the response to growth-promoting estrogens, and renders DNA less vulnerable to damage. According to the leading hypothesis, a full-term pregnancy early in life protects against breast cancer precisely because it reduces a woman's vulnerability to carcinogens and other cancer promoters, such as estrogens.

One of the principal proponents of this hypothesis, the Harvard epidemiologist Nancy Krieger, has urged its further testing. She has also urged a redirection of breast cancer research toward environmental questions.[26] Investigators have repeatedly confirmed that reproductive history contributes to breast cancer risk. We need to know now, Krieger argues, whether women with similar reproductive histories but divergent exposure to carcinogens have marked differences in breast cancer incidence. This need is made urgent by the results of animal studies showing that exposure to certain organochlorines hastens the onset of puberty.[27] Early first menstruation—along with late parenthood—is considered a risk factor for breast cancer in women.

Within the scientific community, grand arguments have ensued from the attempt to classify and quantify cancer deaths due to specific causes.[28] Traditionally, the final result of this task takes the visual form of a great cancer pie sliced to depict the relative importance of different risk factors. "Smoking" is always a big wedge, monopolizing about 30 percent of the circle. "Diet" is also a sizable helping. Depending on who's doing the apportioning, an array of other lifestyle factors—"alcohol," "reproductive and sexual behavior," and "sedentary way of life"—make up the remainder, along with "occupation" and "pollution."

The quarreling begins immediately. How do we account for malignancies, such as certain liver cancers, to which both drinking and job hazards contribute? Or lung and bladder cancers where both job hazards and smoking conspire? Should the effects of pesticides be tallied under "pollution" or under "diet"? What about pollution's indirect effects—such as hormonal disruption, inhibition of apoptosis (programmed death of damaged cells), and immune system suppression that act to augment the dangers of risk factors across the board? What about formaldehyde, which seems to bind with DNA in such a way that it prevents repair of damage induced by ionizing radiation, possibly raising the cancer risk from medical X-rays? Interactions between risk factors aside, how can the death toll from environmental factors be calculated at all when the vast majority of industrial chemicals in commerce have never been tested for their ability to cause cancer?

The futility of what the cancer historian Robert Proctor calls "the percentages game"[29] has not deterred public health agencies from using this kind of simplistic accounting to formulate cancer control policies and educational programs. Lifestyle is the bull's-eye of cancer prevention efforts, while targeting environmental factors, perceived as making a small contribution to the cancer problem, is seen as inefficient.[30] Moreover, the rationale continues, not enough is known about environmental risks to make specific recommendations. (On the other hand, incomplete and inconsistent evidence about the role of dietary fat in contributing to breast cancer does not appear to be an obstacle to advising women to change their diets.)

In my own home state, a recent county-by-county cancer report reproduced an old cancer pie chart published in 1981 that relegated environmental factors to a single, tiny slice and depicted tobacco and diet as major risk factors. The report concluded, "Many persons could reduce their chances of developing or dying from cancer by adopting healthier lifestyles and by visiting their physicians regularly for cancer-related checkups."[31] It never mentions or considers that Illinois is a leading producer of hazardous waste, a heavy user of pesticides, and home to an above-average number of Superfund sites. Nor does this report correlate cancer statistics with Toxics Release Inventory data or attempt to determine whether cancer might follow industrial river valleys, rise in areas of high pesticide use, or cluster around contaminated wells.

Lifestyle and the environment are *not* independent categories that can be untwisted from each other: To talk about one is to talk about the other. A discussion about dietary habits is necessarily also a discussion about the food chain. To converse about childbirth and breast cancer is also to converse about changing the susceptibility to carcinogens in the breast. And to advise those of us at risk for bladder cancer to "void frequently" is to acknowledge the presence of carcinogens in the fluids passing through our bodies.

The Right to Know

During the last year of her life, Rachel Carson discussed before a U.S. Senate subcommittee her emerging ideas about the relationship between environmental contamination and human rights.[32] She urged recognition of an individual's right to know about poisons introduced into one's environment by others and the right to protection against them. These ideas are Carson's final legacy.[33]

The process of exploration that results from asserting our right to know about carcinogens in our environment is a different journey for every person. For all of us, however, I believe it necessarily entails a three-part inquiry. Like the Dickens character Ebenezer Scrooge, we must first look back at our past, then reassess our present situation, and finally summon the courage to imagine an alternative future.

We must begin retrospectively for two reasons. First, we carry in our bodies many carcinogens that are no longer produced and used domestically, but which linger in the environment and in human tissue. Appreciating how even today we remain in contact with banned chemicals such as polychlorinated biphenyls (PCBs) and DDT requires a historical understanding. Second, because cancer is a multicausal disease that unfolds over a period of decades, exposures during young adulthood, adolescence, childhood—and even prior to birth—are relevant to our present cancer risks. We need to discover what pesticides were sprayed in our neighborhoods and what sorts of household chemicals our parents stored under the kitchen sink. Reminiscing with neighbors, family members, and elders in the community where one grew up can be an eye-opening first step.

This part of the journey is, in essence, a search for our ecological roots. Just as awareness of our genealogical roots offers us a sense of heritage and cultural identity, our ecological roots provide a particular appreciation of who we are biologically. It means asking questions about the physical environment we have grown up in and the molecules of which are woven together with the strands of DNA inherited from our genetic ancestors. After all, except for the original blueprint of our chromosomes, all the material that is us—from bone to blood to breast tissue—has come to us from the environment.

Going in search of our ecological roots has both intimate and far-flung dimensions. It means learning about the sources of our drinking water (past and present), about the prevailing winds that blow through our communities, and about the agricultural system that provides us food. It involves visiting grain fields, as well as cattle lots, orchards, pastures, and dairy farms. It demands curiosity about how pests in our apartment buildings are exterminated, how our clothing is cleaned, and how golf courses are maintained. It means asserting our right to know about any and all toxic ingredients in such products as household cleaners, paints, and cosmetics. It requires a determination to discover the location of underground storage tanks, how the land was used before a subdivision was built over it, what is being sprayed along the roadsides and rights-of-way, and what exactly goes on behind that barbed-wire fence at the end of the street. Acquiring a copy of the Toxics Release Inventory for one's home county, as well as a list of local hazardous waste sites is a simple place to begin.

In full possession of our ecological roots, we can begin to survey our present situation. This requires a human rights approach. Such an approach recognizes that the current system of regulating the use, release, and disposal of known and suspected carcinogens—rather than preventing their generation in the first place—is intolerable. So is the decision to allow untested chemicals free access to our bodies until they are finally assessed for carcinogenic properties. Both practices show reckless disregard for human life.

A human rights approach would also recognize that we do not all bear equal risks when carcinogens are allowed to circulate within our environment.[34] Workers who manufacture carcinogens are exposed to higher levels, as are those who live near the chemical graveyards that serve as their final resting place. Moreover, people are not uniformly vulnerable to the effects of environmental carcinogens. Individuals with genetic predispositions, infants whose detoxifying mechanisms are not yet fully developed, and those with significant prior exposures may all be affected more profoundly. Cancer may be a lottery, but each of us does not hold equal chances of "winning." When carcinogens are deliberately or accidentally introduced into the environment, some number of vulnerable persons are consigned to death. The impossibility of tabulating an exact body count does not alter this fact. A human rights approach to cancer strives, nonetheless, to make these deaths visible.

Suppose we assume for a moment that the most conservative estimate concerning the proportion of cancer deaths due to environmental causes is absolutely accurate. This estimate, put forth by those who dismiss environmental carcinogens as negligible, is 2 percent.[35] Though others have placed this number far higher,[36] let's assume for the sake of argument that this lowest value is absolutely correct. Two percent means that 10,940 people in the United States die each year from environmentally caused cancers.[37] This is more than the number of women who die each year from hereditary breast cancer—an issue that has launched multi-million dollar research initiatives. This is more than the number of children and teenagers killed each year by firearms—an issue that is considered a matter of national shame. It is more than three times the number of nonsmokers estimated to die each year of lung cancer caused by exposure to secondhand smoke—a problem so serious it warranted sweeping changes in laws governing air quality in public spaces. It is the annual equivalent of wiping out a small city. It is thirty funerals every day.

None of these 10,940 Americans will die quick, painless deaths. They will be amputated, irradiated, and dosed with chemotherapy. They will expire privately in hospitals and hospices and be buried quietly. Photographs of their bodies will not appear in newspapers. We will not know who most of them are. Their anonymity, however, does not moderate this violence. These deaths are a form of homicide.[38]

According to the most recent tally, forty possible carcinogens appear in drinking water, sixty are released by industry into ambient air, and sixty-six are routinely sprayed on food crops as pesticides.[39] Whatever our past exposures, this is our current situation.

Guiding Principles for Reducing Toxics

After having carefully appraised the risks and losses that we have endured by tolerating this situation, we can begin to imagine a future in which our right to an environment free of such substances is respected. It is unlikely that we will ever rid our environment of all chemical carcinogens. However, as Rachel Carson herself observed, the elimination of a great number of them would reduce the carcinogenic burden we all bear and thus would prevent considerable suffering and loss of human life.[40] Three key principles can assist us in this effort.

One is the idea that public and private interests should act to prevent harm before it occurs. This is known as the *precautionary principle,* and it dictates that *indication* of harm, rather than *proof* of harm, should be the trigger for action—especially if delay might cause irreparable damage.[41] Central to the precautionary principle is the recognition that we have an obligation to protect human life. Our current methods of regulation, by contrast, appear governed by what some frustrated policymakers have called "the dead body approach": Wait until damage is proven before taking action.[42] It is a system tantamount to running an uncontrolled experiment using human subjects.

Closely related to the precautionary principle is the *principle of reverse onus.*[43] According to this edict, it is safety, rather than harm, that should necessitate demonstration. This reversal essentially shifts the burden of proof from the public to those who produce, import, or use the substance in question. The principle of reverse onus requires that those who seek to introduce chemicals into our environment first show that what they propose to do is almost certainly not going to hurt anyone. This is already the standard we uphold for pharmaceuticals and yet for most industrial chemicals, no firm requirement for advance demonstration of safety exists. Chemicals are not citizens. They should not be presumed innocent unless proven guilty, especially when a verdict of guilt requires some of us to sicken and die in order to provide the necessary evidence.

Finally, all activities with potential public health consequences should be guided by the *principle of the least toxic alternative,* which presumes that toxic substances will not be used as long as there is another way of accomplishing the task.[44] This means choosing the least harmful way of solving problems—whether it be ridding fields of weeds, school cafeterias of cockroaches, dogs of fleas, woolens of stains, or drinking water of pathogens. Biologist Mary O'Brien advocates a system of assessment of alternatives in which facilities regularly evaluate the availability of alternatives to the use and release of toxic chemicals. Any departure from zero should be preceded by a finding of necessity. These efforts, in turn, should be coordinated with active attempts to develop and make available affordable, nontoxic alternatives for currently toxic processes and with systems of support for those making the transition—whether farmer, corner dry cleaner, hospital, or machine shop. The highest priority for transformation should be assigned to all processes that generate dioxin or require the use or release of any known human carcinogen such as benzene and vinyl chloride.

The principle of the least toxic alternative would move us away from protracted, unwinnable debates over how to quantify the cancer risks from each carcinogen released into the environment and where to set legal maximum limits for their presence in air, food, water, the workplace, and consumer goods. As O'Brien observed, "Our society proceeds on the assumption that toxic substances will be used and the only question is how much. Under the current system, toxic chemicals are used, discharged, incinerated, and buried without ever requiring a finding that these activities are necessary" (personal communication, M. O'Brien, 1997). The principle of the least toxic alternative looks toward the day when the availability of safer choices makes the deliberate and routine release of chemical carcinogens into the environment as reprehensible as the practice of slavery.

Notes

1. R. A. Weinberg, "A Molecular Basis of Cancer," *Scientific American* (November 1983):126–42.

2. I. Orlow et al., "Deletion of the p16 and p15 Genes in Human Bladder Tumors," *Journal of the National Cancer Institute* 87 (1995):1524–29; S. H. Kroft and R. Oyasu, "Urinary Bladder Cancer: Mechanisms of Development and Progression," *Laboratory Investigation* 71 (1994):158–74; P. Lipponen and M. Eskelinen, "Expression of Epidermal Growth Factor Receptor in Bladder Cancer as Related to Established Prognostic Factors, Oncoprotein Expression and Long-Term Prognosis," *British Journal of Cancer* 69 (1994):1120–25.

3. D. Lin et al., "Analysis of 4-Aminobiphenyl-DNA Adducts in Human Urinary Bladder and Lung by Alkaline Hydrolysis and Negative Ion Gas Chromatography-Mass Spectrometry," *Environmental Health Perspectives* 102 (Suppl. 6) (1994):11–16; P. L. Skipper and S. R. Tannenbaum, "Molecular Dosimetry of Aromatic Amines in Human Populations," *Environmental Health Perspectives* 102 (Suppl. 6) (1994):17–21; S. M. Cohen and L. B. Ellwein, *Environmental Health Perspectives* 101 (Suppl. 5) (1994):111–14.

4. P. Vineis and G. Ronco, "Interindividual Variation in Carcinogen Metabolism and Bladder Cancer Risk, "*Environmental Health Perspective* 98 (1992):95–99.

5. One researcher offers the following reflection on the bladder cancer situation in England: "The continued use of known carcinogenic substances in British industry for many years after their identification, the wide range of industries with a known or suspected increased risk of bladder cancer, and our ignorance of the carcinogenic potential of many materials used in current manufacturing should be a cause for continuing concern" (R. R. Hall, "Superficial Bladder Cancer," *British Medical Journal* [1994]:910–13).

6. D. T. Silverman, "Urinary Bladder," in *Cancer Risks and Rates,* NIH Pub. 96-691, A. Harras, ed. (Bethesda, Md.: National Cancer Institute, 1996, pp. 197–99). Routine screening for bladder cancer is not done. Thus earlier detection or improved diagnostic techniques are unlikely explanations for the recent increases in rates. R. A. Schulte et al. (eds.), "Bladder Cancer Screening in High Risk Groups," *Journal of Occupational Medicine* 32 (1990):787–945.

7. Environmental Protection Agency, *1992 Toxic Chemicals Release Inventory: Public Data Release.* EPA 745-R-001 (Washington, D.C.: EPA, 1994, p. 79).

8. U.S. Department of Health and Human Services, *Seventh Annual Report on Carcinogens* (Research Triangle Park, N.C.: USDHHS, 1994, p. 389).

9. E. M. Ward et al., "Monitoring of Aromatic Amine Exposure in Workers at a Chemical Plant with a Known Bladder Cancer Excess," *Journal of the National Cancer Institute* 88 (1996):1046–52.

10. R. Ouellet-Hellstromt and J. D. Rench, "Bladder Cancer Incidence in Arylamine Workers," *Journal of Occupational and Environmental Medicine* 38 (1996):1239–47; J. D. Rench et al., *Cancer Incidence Study of Workers Handling*

Mono- and Di-arylamines Including Dichlorobenzidine, ortho-Toluidine, and ortho-Dianisidine (Falls Church, Va.: SRA Technologies, 1995); "Study Finds Bladder Cancer Threat Among Conn. Plant Workers," *Boston Globe,* September 21, 1995, p. 42.

11. Francis Collins, Richard Klausner, and Kenneth Olden, statement on cancer, genetics, and the environment before the Senate Committee on Labor and Human Resources, March 6, 1996 (U.S. Department of Health and Human Services press release).

12. National Cancer Institute, *Understanding Gene Testing,* NIH Pub. 96-3905 (Bethesda, Md.: NCI, 1995).

13. G. Marra and C. R. Boland, "Hereditary Nonpolyposis Colorectal Cancer: The Syndrome, the Genes, and Historical Perspectives," *Journal of the National Cancer Institute* 87 (1995):1114–25; N. Papadopoulos et al., "Mutation of a *mutL* Homolog in Hereditary Colon Cancer," *Science* 263 (1994):1625–29.

14. Bert Vogelstein, "Heredity and Environment in a Common Human Cancer," lecture at Harvard Univ. Medical School, May 3, 1995). In exploring the use of the term "sporadic" by cancer researchers, historian Robert Proctor observed, "The presumption is apparently that heredity is orderly, while environmental causation is chaotic, perhaps even indecipherable. . . . Genetics offers hope for new forms of therapy, but also seems to imply resignation with regard to the possibility of prevention." See R. N. Proctor, *Cancer Wars: How Politics Shapes What We Know and Don't Know About Cancer* (New York: Basic Books, 1995, p. 245).

15. Five to 10 percent is the estimate most often cited. A recent prospective cohort study of more than 100,000 women placed this figure even lower—at about 2.5 percent. See G. A. Colditz, "Family History, Age, and Risk of Breast Cancer: Prospective Data from the Nurses' Health Study," *Journal of the American Medical Association* 2 (70) (1993):338–43.

16. D. Holzman, "Mismatch Repair Genes Matched to Several New Roles in Cancer," *Journal of the National Cancer Institute* 88 (1996):950–51.

17. "Cancer Prevention" (pamphlet) (Bethesda, Md.: U.S. Department of Health and Human Services, n.d.).

18. G. Edlin, *Human Genetics. A Modern Synthesis,* 2d ed. (Boston: Jones & Bartlett, 1990). Quotations are from pages 184–204.

19. C. E. Rosenberg, *The Cholera Years: The United States in 1832, 1849, and 1866* (Chicago: Univ. of Chicago Press, 1962, pp. 1–60).

20. Some researchers argue that "delayed childbirth" among white women explains much of the elevated incidence of breast cancer in the northeastern states. See S. R. Sturgeon, "Geographic Variation in Mortality from Breast Cancer among White Women in the United States," *Journal of the National Cancer Institute* 87 (1995):1846–53.

21. M. P. Madigan, "Proportion of Breast Cancer Cases in the United States Explained by Well-Established Risk Factors," *Journal of the National Cancer Institute* 87 (1995):1681–85.

22. D. J. Hunter et al., "Cohort Studies of Fat Intake and the Risk of Breast Cancer—A Pooled Analysis," *New England Journal of Medicine* **334** (1996): 356–61;·D. J. Hunter and W. C. Willett, "Diet, Body Size, and Breast Cancer," *Epidemiology Reviews* **15** (1993):110–32; E. Giovannucci et al., "A Comparison of Prospective and Retrospective Assessments of Diet in the Study of Breast Cancer," *American Journal of Epidemiology* **137** (1993):502–11. The role of dietary fat in creating breast cancer risk remains uncertain in part because the range of fat intake among the various groups of women studied has so far been relatively narrow.

23. As two leading researchers have observed, energy intake from fat has been declining as breast cancer has increased: Hunter and Willett, "Diet, Body."

24. Drs. Devra Lee Davis, Samuel Epstein, and Janette Sherman are among the researchers calling for a more ecological approach to diet. See S. S. Epstein, "Environmental and Occupational Pollutants Are Avoidable Causes of Breast Cancer," *International Journal of Health Services* **24** (1994):145–50; and J. Sherman, *Chemical Exposure and Disease: Diagnostic and Investigative Techniques* (Princeton, N.J.: Princeton Scientific Publishing, 1994, p. 83).

25. Consumption of animal fat (or meat) is most strongly linked to colon and prostate cancers. See W. C. Willett, "Diet and Nutrition," in *Cancer Epidemiology and Prevention,* 2d ed. D. Schottenfeld and J. F. Fraumeni, Jr., eds. (Oxford: Oxford Univ. Press, 1996, pp. 438–61).

26. N. Krieger, "Exposure, Susceptibility, and Breast Cancer Risk," *Breast Cancer Research and Treatment* **13** (1989):205–23.

27. This topic is currently under exploration by Dr. Mary Wolff, who is interested in all factors, including childhood diet and level of physical activity, that contribute to the onset of puberty in girls. M. S. Wolff, "Organochlorines and Breast Cancers," presentation at the American Public Health Association, New York, November 20, 1966. See L. M. Walters et al., "Purified Methoxychlor Stimulates the Reproductive Tract in Immature Female Mice," *Reproductive Toxicology* **7** (1993):599–606; P. L. Whitten et al., "A Phytoestrogen Diet Induces the Premature Anovulatory Syndrome in Lactionally Exposed Female Rats," *Biology of Reproduction* **49** (1993):1117–21; R. J. Gellert, "Uterotropic Activity of Polychlorinated Biphenyls and Induction of Precocious Reproductive Aging in Neonatally Treated Female Rats," *Environmental Research* **16** (1978): 123–30.

28. See, for example, R. Doll and R. Peto, *The Causes of Cancer: Quantitative Estimates of Avoidable Risks of Cancer in the United States Today* (Oxford: Oxford Univ. Press, 1981); and a rebuttal by S. S. Epstein and J. B. Swartz, "Fallacies of Lifestyle Cancer Theories," *Nature* **2**(89) (1981):127–30.

29. Described in Proctor, *Cancer Wars,* pp. 54–74. See also J. M. Kaidor and K. A. L'Abbe, "Interaction between Human Carcinogens," in *Complex Mixtures and Cancer Risk,* H. Vainio et al., eds., IARC Scientific Pub. 104 (Lyon, France: International Agency for Research on Cancer, 1990, pp. 35–43).

30. The American Cancer Society does not discuss environmental factors in its recent report on cancer prevention. See American Cancer Society, *Cancer Risk Report: Prevention and Control,* 1995 (Atlanta Ga.: ACS, 1995). See also K. R. McLeroy, "An Ecological Perspective on Health Promotion Programs," *Health Education Quarterly* **15** (1988):351–77.

31. Illinois Department of Public Health, *Cancer Incidence in Illinois by County, 1985–87,* Supplemental Report (Springfield, Ill.: IDPH, 1990, pp. 7–8).

32. Rachel Carson on environmental human rights: Senate testimony hearings before the Subcommittee on Reorganization and International Organizations of the Committee on Government Operations, "Interagency Coordination in Environmental Hazards (Pesticides)," U.S. Senate, 88th Cong., 1st sess., June 4, 1962.

33. Carson, *Silent Spring* (Boston, Mass.: Houghton Mifflin, 1962, pp. 277–78).

34. R. Perera, "Uncovering New Clues to Cancer Risk," *Scientific American* (May 1996): 54–62; S. Venitt, "Mechanisms of Carcinogenesis and Individual Susceptibility to Cancer, *Clinical Chemistry* **40** (1994):1421–25; G. W Lucier, "Not Your Average Joe" (editorial), *Environmental Health Perspectives* **103** (1995):10.

35. Harvard Center for Cancer Prevention, "Harvard Report on Cancer Prevention," *Cancer Causes and Control* 7 (Suppl. 1) (1996):3–59; D. Trichopoulos et al., "What Causes Cancer?" *Scientific American* (September 1996):80–87.

36. Proctor, *Cancer Wars.*

37. 10,940 is 2 percent of 547,000, the projected figure for total cancer deaths in 1995. See American Cancer Society, *Cancer Facts and Figures—1995,* rev. (Atlanta, Ga.: ACS, 1995).

38. The environmental analysts Paul Merrell and Carol Van Strum have argued that the concept of acceptable risk is tolerated only because of the anonymity of its intended victims. See P. Merrell and C. Van Strum, "Negligible Risk: Premeditated Murder?" *Journal of Pesticide Reform* 10 (1990):20–22. Likewise, the molecular biologist and physician John Gofman has argued, "If you pollute when you DO NOT KNOW if there is any safe dose (threshold), you are performing improper experimentation on people without their informed consent. . . . If you pollute when you DO KNOW that there is no safe dose with respect to causing extra cases of deadly cancers, then you are committing premeditated random murder" (J. W. Gofman, memorandum to the U.S. Nuclear Regulatory Commission, May 21, 1994).

39. M. Eubanks, "Biomarkers: The Clues to Genetic Susceptibility," *Environmental Health Perspectives* **102** (1994):50–56.

40. Carson, *Silent Spring,* p. 248. See also M. J. Kane, "Promoting Political Rights to Protect the Environment," *Yale Journal of International Law* **18** (1993): 389–411.

41. This principle was endorsed in 1987 by European environmental ministers in a meeting about the deterioration of the North Sea. [K. Geiser, "The Greening

of Industry: Making the Transition to a Sustainable Economy," *Technology Review* (August/September 1991):65–72.] See also T. O'Riordan and J. Cameron (eds.), *Interpreting the Precautionary Principle* (London: Earthscan, 1994).

42. Devra Lee Davis, quoted in "Is There Cause for 'Environmental Optimism'?" *Environmental Science and Technology* **29** (1995):366–69.

43. This principle has been embraced by the International Joint Commission in their Eighth Biennial Report on Great Lakes Water Quality (Washington, D.C., and Ottawa, Ontario: International Joint Commission, 1996, pp. 15–17). See also discussions of proof in T. Colborn et al., *Our Stolen Future: Are We Threatening Our Fertility, Intelligence, and Survival?—A Scientific Detective Story* (New York: Dutton, 1996); and G. K. Durnil, *The Making of a Conservative Environmentalist: With Reflection on Government, Industry, Scientists, the Media, Education, Economic Growth, and the Sunsetting of Toxic Chemicals* (Bloomington: Indiana Univ. Press, 1995).

44. My ideas on this topic are inspired in part by those of biologist Mary O'Brien. See M. H. O'Brien, "Alternatives to Risk Assessment: The Example of Dioxin," *New Solutions: A Journal of Environmental and Health Policy* 3 (Winter 1993): 39–42; and K. Geiser, "Protecting Reproductive Health and the Environment: Toxics Use Reduction," *Environmental Health Perspectives* **101** (Suppl. 2) (1993):221–25.

3

Deconstructing Standards, Reconstructing Worker Health

Charles Levenstein and John Wooding

An 18-year-old Vietnamese émigré to the United States, working for a cleaning and maintenance company, is crushed in a printing plant when a press that he has been cleaning malfunctions and he goes unwittingly to his death. A 35-year-old secretary—a white woman—is forced to undergo operations on both of her wrists because of repetitive strain injuries she has incurred at her computer workstation. A 60-year-old African American, a "retired" miner, sits in a rocker on his front porch carefully gasping for breath, hoping his heart holds out under the strain of oxygen deprivation.

Worker disease and injury flow directly out of the choices of technology made by employers. The use of labor; the intensity of the work; the machines, workstations, and chemicals that endanger worker health—the "working conditions"—are determined by managers, company planners, engineers, and sometimes corporate lawyers whose minds focus on one thing only—making profit.

Occupational disease and injury are "unintended consequences" of technological choices driven by financial imperatives: That is what we mean when we say that the social relations of production determine the health and well-being of workers. Identifying the problem, however, is easier than finding a solution to it. If workers had strong unions and exerted substantial political power in the country, they might be able to play a serious role in the evaluation of the technologies of production, and even prevent the use of dangerous processes and substances.

What, however, would they do? Demand better safety standards? Limit uncontrolled exposures to hazardous chemicals? Legislate reasonable hours of work? Perhaps worker organizations would assert control over what technologies to use and search out ways that work could enhance life, develop worker talents, teach cooperation, and produce goods that were truly socially useful. Which would be better, to demand improved standards or to assert democratic control over technology?

On the other hand, suppose that workers have little power and their organizations are small and/or limited. Suppose that they are embattled and defensive, constantly on the alert for assaults by employers who want union-free environments. Suppose that the best they can imagine is a slightly higher wage and a little bit of job security. Suppose that they can fight over only the most life-threatening situations and have learned to accept chronic hazards; suppose too that they are thankful to have jobs at all, much less ones that are safe.

Under such conditions, what are reasonable demands? How can workers engage in the struggle against toxic chemicals? Who are the allies that can help workers achieve some modicum of decency and safety? In a situation in which workplace democracy seems a distant dream, should workers fight for occupational health standards that are at least minimally protective? The solutions to workplace hazards in each of these polar situations are all quite different. The answers must be appropriate to the political setting, to the political possibilities. To speak of "worker control" in the United States or Malaysia, where the numbers of private sector unions are about the same, is quite different from discussing the limits of expert-dominated standard setting in Scandinavia, where the unions and their political allies are quite powerful.

This chapter focuses on the contradictory nature of occupational health standards in the United States—a country with a weak labor movement; a deep ideological commitment to science and progress; and a largely hidden problem with occupational injury, disease, and death. The struggle for environmental health begins, we believe, with the effort to control toxic exposure in the workplace. What is produced at the point of manufacture ultimately becomes the potential source of environmental hazard.

The Debate over Standards

Because the approach to controlling occupational hazards in the United States consists of the enforcement by the state of standards, occupational health professionals spend a great deal of time developing objective measures of exposures and health effects. A serious debate has emerged, however, about the usefulness of such measures. Critics such as Eileen Sen Tarlau argue that the focus on precise environmental measurements misdirects attention from struggles in the work environment to highly technical arguments over standards derived from scientific assessments of contaminants.[1] Indeed, in most workplaces it is difficult to find a lack of compliance with these standards for chemicals. Similarly, in the environmental movement, the stress on restricting or banning chemicals creates a kind of chemical fetishism that fails to deal with the underlying imbalance of power among managers, workers, and citizens. That is, it fails to probe the underlying social relations involved.

Using Marx's discussion of commodity fetishism as a starting point, this chapter argues that chemicals as "commodities" have very real power: the power to maim, destroy, and kill. The way labor is used is a product of the social relations of production; the use of potentially dangerous materials is a function of the economic priorities of a firm. Thus the production and use of chemicals are products of economic and social forces. To focus solely on chemicals (as many in the worker health and safety and environmental movements have) obscures the underlying power relations in the workplace. The struggle for environmental health is a struggle over political power.

Making chemicals a fetish in this way obscures the options available for exploring alternative forms of production. As Marx was aware, mature capitalist economies create things that are not themselves commodities, but which are perceived as such. The underlying human relations become reified.[2] Such, we argue, has been the fate of many chemical contaminants and standards. In this respect, the chemical "enemy" becomes the central concern, not the system that produces the chemical, or the social and political relations that enable it to be produced and used.

The narrow focus on chemicals has a tendency to play into the hands of the most powerful actors in the struggle over health and safety. As a consequence, very large corporate interests frequently respond to concerns about specific chemicals (or calls to ban them) by lobbying and making claims about generalized costs and benefits to society, about loss of jobs and the maintenance of standards of living. For instance, the American Petroleum Institute delayed adoption of a more stringent benzene standard in the 1980s, insisting that the Occupational Health and Safety Administration (OSHA) demonstrate a significant improvement in health. According to one analyst, this delay caused a significant number of deaths while OSHA accumulated the additional evidence.[3]

Movements that challenge these claims confront enormous social and economic power, and typically do not win the fight. The focus on the chemical weakens the articulation of real resistance to environmental threats. On the other hand, organizers claim that it is far easier to mobilize workers and/or citizens around concerns about particular toxic substances rather than more abstract systemic critiques. This is not to deny that particular substances are usually of serious concern and that even small victories may save many lives.

Much the same story is true for the way we perceive the role that standard setting plays in the struggle for a healthier workplace. Despite knowledge about the biases in state and federal regulation, despite cynicism about the regulatory process, and despite the evidence for lack of enforcement, we still use the "standard" as a gospel. It, too, becomes reified. It too, becomes a Holy Grail, to be pursued. Why is this?

We consider this problem by analyzing the ways in which some Left critics have tended to make fetishes of chemicals, focusing on the ways in which toxic substances harm workers rather than on the production systems that create these chemicals.[4] This fetishization, in turn, translates into a heavy reliance on regulatory standards to protect workers and the environment. The focus on a particular hazard draws the struggle for a healthier society away from the underlying social and economic inequities that are at the root of deteriorating workplace and environmental conditions. Often it results in a prolonged scientific and legal debate that exhausts the resources of workers and their unions, environmentalists, and community groups. This creates significant problems in the struggle for environmental, worker, and community health. It suggests that we should pay greater attention to the politics of production (who makes the decision to produce a certain product and in what way, who controls issues of safety and health, etc.) and strategies for reform, rather than to specific chemicals, if we are to clean up the work environment and create momentum for long-run political change.

What Is a Standard?

The Legal Terrain
In the United States, federal standards to protect worker health and safety were established with the passage of the Occupational Safety and Health Act in 1970. In order to give OSHA's compliance officers regulations to enforce, a round of consensus standards were adopted en masse: these were "voluntary" standards developed by the American National Standards Institute, the American Conference of Governmental Industrial Hygienists, and other private standard-setting groups. In addition, OSHA was given a procedure for establishing new standards. This procedure included public hearings, publication of proposed standards, promulgation of new standards, and judicial review. In the case of newly discovered but inadequately regulated hazards, OSHA had the authority to issue emergency temporary standards. Essentially, the federal enforcement of these standards was the nature of the intervention envisaged by the act.[5]

This is in contrast to the occupational health regulatory regimes of other countries, such as Italy and those in the United Kingdom. In countries such as these, labor organizations have a much larger role in the negotiation of standards and their enforcement; labor has the right to act to protect itself through regulations.[6] In the United States, labor had (and has) the right to complain to the government, but does not have the authority to enforce regulations on the shop floor. Indeed, in the early 1970s some (dissident) Americans referred to the Occupational Health and Safety Act as the "full employment act for industrial hygienists," reflecting the power of professionals, but not that of labor.

Many people think of standards as "permissible exposure limits"—a secret code of parts per million kept in large but obscure handbooks owned and controlled by industrial hygienists and OSHA inspectors. Some standards are like that. However, the more elaborate ones, developed over OSHA's 28-year history, may include requirements for medical surveillance and record maintenance, worker training and worker removal when appropriate, and "action" levels that kick in well in advance of violation of permissible exposures. Specific standards are in force concerning medical records and communication about hazards. For hazardous waste workers and emergency responders, Congress mandated that OSHA write a special standard that includes medical surveillance and worker training.[7]

By and large these standards have been written according to basic principles of industrial hygiene. They reflect the preference for engineering controls rather than personal protective equipment and behavioral change. They are, of course, conditioned by the legislative requirements that the standards be "feasible" and the presidential (and Supreme Court)-imposed requirement that regulations meet a variety of cost-benefit criteria.

The most controversial standard proposed by OSHA has been the draft ergonomics standard, dealing with materials handling, the positioning and design of workstations, and in general the relationship between the worker and his or her physical environment in the workplace. While the chemical regulations have affected relatively small groups of businesses, virtually all firms view the ergonomics standard as a threat since ergonomic hazards are not rare, and injuries caused by poor work design in materials handling are frequent. A broad-based business coalition led by United Parcel Service and Liberty Mutual Insurance has effectively brought OSHA's ergonomics initiative to a halt.

Perhaps what is most intriguing about battles over standard setting is the presumption that the government is capable and desirous of enforcing them. From the earliest days of OSHA, it was plain that the agency would be underfunded and understaffed. The law provided for state assumption of enforcement activities with the understanding that state efforts would be at least as effective as federal ones. In particular, during the Reagan and Bush administrations, delegating enforcement activities to the states meant the undermining of standards enforcement.

The Political Economy of Standards

Under what conditions can we expect a standards approach to intervention to be effective in protecting worker health and safety? Weeks, for example, has demonstrated that the regulatory regime of the Mine Safety and Health Administration, before the deregulation and "reform" of the

1980s, was successful in improving the health of mineworkers. Inspectors were well-trained, unions were consulted, and standards were enforced.[8]

What would "good" standards-based regulation look like? First, the development of standards presumes and requires that the government will enforce them. Second, it assumes that management will comply with the standards and accept the regulation of the work environment. Finally, a strong, knowledgeable, and committed trade union completes the triangle.

On the other hand, an approach based on health standards cannot be successful with a government committed to deregulation, an industry faced with declining or highly competitive markets, unorganized workplaces, and weak unions. This is the major problem we face today. Creating standards based on the assessment of the health impact of a chemical or the dangers inherent in a process is a necessary but not sufficient condition for effective regulation.

By 1970, when the Occupational Health and Safety Act was enacted, the United States confronted a new and highly competitive world economy in which American goods and American companies no longer dominated. In addition, U.S. and European-based multinational corporations began to spread their activities across the globe, setting up production facilities in many developed and developing countries. These multinational corporations invested heavily abroad, seeking new markets and new places to produce with lower wages, less regulation, and less taxation. Aided by new communications systems and new opportunities for investment, industry and its accompanying investment capital have become increasingly mobile. This situation undercuts the ability of advanced industrial countries to regulate domestic industry for fear that their industries might flee regulation. At the same time, it spreads advanced technological and other hazards to countries that do not have the social or scientific infrastructures to protect their citizens and to others that are prevented from banning hazardous materials by trade agreements.[9]

These developments have had a negative impact on workers in the United States: real wages have fallen and housing, education, and medical costs have all increased. Despite more two-earner families, American workers are worse off now than they were in 1970.

Although the United States is in a period of relatively full employment, we continue to wrestle not only with a refractory unemployment problem but also with other difficult problems for workers created by changes in the structure of the economy. For example, the manufacturing industry in the United States, while remarkably productive, has been declining as a source of employment. Some workers, forced out of relatively high-paying unionized jobs in manufacturing, have had to take minimum wage employment in the service industry. Years of accumulated skills and experiences have evaporated and with them, the middle-class lifestyle that manufacturing employment supported.

The service industry includes a wide array of firms, from fast-food outlets to high-tech consulting companies.[10] Much of the service industry is unorganized, so the labor movement suffers as its manufacturing bulwark declines and workers move into small nonunion firms with low pay, few benefits, and few rights.

These profound changes in the economy have undercut the government's willingness to regulate firms for fear of damaging their competitiveness, and they have undercut our ability to demand effective workplace and environmental regulation. The large manufacturing firm has become a dinosaur. Voluntary compliance in manufacturing rested on professionally trained health and safety departments, but these are declining as firms decentralize or contract out their operations. Most workers are in small, non-union firms that do not employ work environment professionals. In this social context, standards-based regulatory approaches are unlikely to be effective.

The Politics of Regulation

The politics of regulation is often considered in relation to the battles between labor and capital, unions and employers, environmentalists and corporate owners, and the lobbyists for both sides. The American model presumes that such regulatory politics flows from the negotiations of democratically elected representatives, using scientific knowledge to determine the most efficient, and occasionally the most health-protecting, regulatory standards. Sometimes it even works like this. However, the system depends on another kind of politics—a cultural politics that reflects the intersection of deeply held beliefs about science and the rule of law.

The Hegemony of Science

For many Left critics, the notion that contemporary science is socially constructed is hardly new. In occupational health science what that means is that social factors are always of considerable importance in shaping the definition of disease and its amelioration.[11] There is a long tradition of progressive critique of scientific objectivity and neutrality.[12]

The French sociologist Pierre Bourdieu argues that the holders of intellectual capital have power over others.[13] Similarly, Aronowitz's discussion of science and power reveals that—at least for the United States—it is science and scientists who possess analogous power.[14] The broad acceptance of science enables and supports the legitimacy of the system of social relations as a whole.

In the field of occupational health and safety, science provides the basis for a limited approach to regulating workplace hazards (what we have characterized as fetishization). Nevertheless, such science reveals real hazards and can provide some solutions that protect workers. What it cannot do, however, is provide a fundamental critique of the system of power relations and the resultant system of production.

Scientific evidence about the toxicity of a chemical, epidemiological evidence about its effect on humans, and clinical evidence about physiological processes conjoin to establish whether a contaminant is guilty of endangering health. Science does not acknowledge a problem if there is no scientific evidence of cause *even if workers get sick or complain of health problems.*

Despite these issues, science has become the foundation on which standards and regulations are built. Progressives and health activists marshal scientific evidence to demand that a chemical be restricted or banned; corporations present scientific evidence that the chemical is innocent, or maybe a little bit guilty. If exposure to a chemical may harm one worker in a hundred, then this may be acceptable given the benefits (investment, jobs, products, taxes, and profits). The question is, therefore, not whether science is "right" or "wrong" but rather, how does political and cultural power get resolved in the struggle over health?

The Hegemony of Legalism and Democracy

As is the case with science, belief in the objectivity and neutrality of the political and judicial system is deeply embedded in American culture. This country was founded supposedly on a "government of laws, not men." However, today democracy founders in the face of corporate control of public policy, the role of big money in the electoral process, the weakness of labor unions as advocates for workers, and the political and social inequalities that engender these antidemocratic tendencies. Yet, belief in the pantheon of liberal rights, and liberal democracy, intersects with broad assumptions about the validity of science and scientific knowledge, and leads advocates to the judicial and political system as a means of protecting worker health and safety. The conjunction of these two deeply held belief systems yields occupational safety and health policy. Or as Noble characterizes it, *Liberalism at Work*.[15]

The standard-setting process is highly controversial and has led to innumerable lawsuits and appeals. Like the inspection system, OSHA standards have been (and are) at the nexus of a political battle about the role of the state in the American political system. The passage and implementation of the health and safety act has created unprecedented political conflicts about the role of government regulation of private enterprise. Clearly, however, the activities of OSHA were (and are) highly circumscribed by American resistance to state intervention in the economy, its vulnerability to political influence, and the overall balance of class forces in the country in the decades after World War II.

These features also explain the genesis of OSHA in the hundred or so years since Massachusetts enacted the first factory inspection law. A liberal-democratic ideology that is antithetical to government intervention (especially at the federal level), a weak and fragmented labor movement, and the absence of a tradition of a professional civil service have all contributed to the slow development of health and safety regulation and the overall weakness of the contemporary system.

These conditions have reduced the regulatory agency's role to a minimum, while at the same time making it lose credibility with both employers and employees. The agency no longer has the authority to command compliance, nor do workers trust it. The result is that OSHA itself has become ineffective. Nevertheless, the OSHA law was vital to labor as the first organized effort to establish a framework with a mandate to meaningfully improve conditions in American workplaces.

The Objectivity of Standards

Definitions of occupational disease, understanding of workplace hazards, indeed, the scientific processes of investigating the work environment itself are made into fetishes. At the microsocial level, we can observe hegemony operating through fetishization. A worker feels sick. He or she will ask, "Is my work making me sick?" They may then ask, "Can I get compensation for my illness, if I have to leave my work? How will I pay my medical expenses? Can I get hazard pay for the work?"

The employer most likely asks two questions: "Is this person really sick? If so, am I liable?" The scientist responds: "Perhaps I can measure a specific quantifiable exposure associated (at a 95 percent level of confidence) with a specific quantifiable health effect. And the lawyers muse: "Is this illness more likely than not the result of this workplace exposure?"

From this multiple of framings emerges a dominant one; thus is "knowledge" created and used as "information." "Aches and pains" are the names that some workers may give to their experience of occupational injury. Ergonomists and other occupational scientists begin to talk about "repetitive strain injuries." A fierce struggle ensues among scientists, workers' compensation insurers and agencies, and relevant counsel, concerning how broadly to construe RSI. And for many, "carpal tunnel syndrome"—a quite narrow construction of "aches and pains"—becomes the disease of choice, although this struggle is ongoing. By way of contrast, through political and economic struggle by their organization of disabled workers, including public demonstrations and trade union reform, U.S. coal miners have been able to establish a broad definition of black lung disease.[16]

At the level of experiencing occupational disease in the workplace, of understanding its nature, and of doing something about it, the politics of science and the politics of regulations merge. All are framed by the faith we have in science and professionalism, legalism and democracy. The system sometimes comes through: The worker is taken seriously, the effect of the contaminant is established, the legal-regulatory apparatus removes the danger. More often than not, however, science cannot definitively establish the relationship between the health problem and the disease. The ill worker is not given credibility. The employer uses science to deny responsibility. The legal system protects the rights of the powerful. Every day this tragic story plays out in thousands of American workplaces.

Alternatives: What Is to Be Done?

While much has been achieved in promoting workplace health and safety by collecting scientific evidence on the toxicity of chemicals, on the physiological hazards of such problems as repetitive-motion injuries, and on the physical dangers inherent in many work processes, we still have a long way to go. When we (1) focus on the scientific evidence for threats to human welfare, (2) believe that the control or removal of the threat will deal with "the problem," and (3) assume that a regulatory process is democratic or that a standard, once established, will protect workers, we fail to see the underlying power relations that determine our agenda.

These power relations, in particular the weakness of organized labor in the United States, are not only of consequence for the health and safety of workers in this country but also have a deep impact on workers abroad. In Europe and the developing world, American scientific evidence and American standards are frequently viewed as the safety benchmarks that set the regulatory agenda for other countries. The regulations are not understood as political phenomena that reflect the power relations in the United States.

We are not Luddites. Science is not the tool of a ruling class, and most scientists make every effort to be objective in what they do. Nevertheless, science and the standards that are derived from it are products of sociopolitical relations that can only, in the long run, benefit those who have power and control in society. Technological progress produces untold benefits to society, but it also creates serious technological hazards (and uncertainty about risk), and myriad social problems.

How, then, do we obtain effective regulation, certain knowledge, and meaningful standards? How could science be used to address the interests of those most affected by chemicals in the workplace, or by processes that threaten health and well being? In short, what would a democratic science look like?

The domination of science at the cultural and practical level requires both lack of knowledge and lack of power on the part of the dominated. Science provides knowledge, but of a particular kind, framed in a particular way. The illusion of power (that we live in a democracy, that government regulations and laws protect us from exploitation, that "we" can shape our choices and chose our rulers) also provides a particular frame for our assumptions about control.

A counterhegemonic approach to setting standards and regulating chemicals must involve, as Antonio Gramsci recognized, a constant "war of maneuver."[17] A crucial step is to constantly expose the bias in science and law, and seek ways to empower workers and communities. What would such activities look like? How can we avoid the endless pieties about worker empowerment and the empty language of "rights?" How can we take all that is good in science and make it appropriate and use it for the protection of worker health and safety and the environment?

First, since the central question is the power relations that exist in society as a whole and in the workplace in particular, it is critical to continue to build organizations that represent the interests of workers. The trade union movement, of course, has been the main vehicle for representing workers. However, in protecting their members from workplace injury and disease, trade unions have been only marginally successful. Certainly they have pressed for protections within collective bargaining agreements, for union health and safety committees, and for increased government regulation. In some cases, they have also developed their own industrial hygiene capacity, and they have also worked with the various coalitions and committees on occupational safety and health (COSH). These grassroots groups of trade union occupational health activists and their supporters advocate and provide technical information, training, and resources to workers and worker organizations. The effectiveness of these activities is increasingly limited, however.

In recent years some trade unions have dismantled much of their internal capacity for dealing with occupational safety and health problems, and the AFL–CIO has reduced its own safety and health staff to a minimum. On the other hand, some building trade unions have substantially increased their commitment to health and safety.

The trade unions have never given their wholehearted commitment to the COSH movement. Indeed, they do not provide them substantial financial support. The unions' other source of scientific and technical support has been the resources commanded by occupational safety and health professionals, scientists, and academicians who are friendly or committed to labor. Historically, these allies of labor have had an ambiguous relationship to much of the trade union movement. Unions tend to be wary of professionals, and the isolation of American trade unionism from genuine political power has left a legacy of suspicion of intellectuals and scientists. The recent emphasis on organizing is understandable given the decline in membership, but health and safety are of great concern to the rank and file and could be a tool for mobilization.

The trade union movement should recommit to occupational safety and health as a central organizing strategy. It should fund COSH groups. It should pursue and develop alliances with scientists, professionals, and intellectuals in academia, and defend regulatory institutions against attack.

Second, the ambivalence of the trade union movement toward intellectuals and scientists raises significant contradictions for those individuals who seek to be allies of labor but who are not themselves in the labor movement. Often caught between the several demands of professional identity, professional ethics, and dependence on corporate or institutional funding, scientists and professionals in the occupational safety and health movement find themselves the victims of numerous contradictory demands. Too often these demands force professionals to view occupational safety and health as a technical, not a political problem. As Lax notes:

professionals clinging to traditional notions of scientific objectivity carried out by a restricted group of experts (professionals) are blind to the ways corporate norms have penetrated knowledge production at all levels. As a result, even professionals who see themselves as worker advocates further the ability of capital to makes its influence invisible as corporate knowledge masquerades as universal truth.[18]

This is not to say that professionals are simply servants of an exploiting and distorting system. Professionals are workers. The demands made on industrial hygienists, scientists, and researchers are similar to those made on traditional blue-collar workers. They need jobs to pay the rent and to eat; they need to be acquiescent to demands placed on them to achieve career advancement; they need to commit to the dominant norms of science to gain legitimacy and thus grant monies and career advancement and to be taken seriously when they do advocate for workers. In short, for professionals to become counterhegemonic they must confront all the contradictions that workers and unions do.

Clearly, professionals must view worker health and safety problems not only in technical terms, accessible to "scientific" solutions, but also as political issues stemming from inequalities in political power. Professionals must also take workers seriously: their demands, their health problems, and their identification of hazardous situations. In addition, professionals should democratize and demythologize science wherever possible, making it understandable and relevant to working people. A new occupational health science must shed the trappings of neutrality and objectivity and reassert its commitment to worker health.

One key to this process lies, not only in changing the attitudes and values of professionals and their associations (which remain largely dominated by corporate interests), but also in the training of those professionals. A counterhegemonic professional education largely depends on the encouragement and support of university and professional school faculty—another key reason why progressives, and trade unions in particular, should build alliances with and support allies within the academic world.

A third and final strategy for challenging the cultural dominance of scientific objectivity, professionalization, and deep dependence on a corrupted standard-setting procedure is to develop systems by which communities can set and enforce their own standards for controlling polluting industries. Building networks of community organizations and trade unions concerned about toxic chemicals, dangerous work, and environmentally threatened neighborhoods is an essential first step toward an alternative to the processes dominated by multinational corporations. In short, we need a democratic standards organization.

These suggestions provide the possibility for exposing the fallacy of scientific objectivity and the inviolable dependence on standards. They are the first steps in a process that could lead to credible protection for workers, not only in the United States, but also in the many countries that use our health and safety standards as a model.

Acknowledgments

The authors would like to thank Greg Delaurier, Richard Hofrichter, and David Kriebel for their very helpful comments on an earlier draft of this paper.

Notes

1. Eileen Senn Tarlau, "Playing Industrial Hygiene to Win," *New Solutions: A Journal of Environmental and Occupational Health Policy,* 1(4) (Spring 1991): 72–81. John Wooding and Charles Levenstein (eds.), *Work, Health and Environment: Old Problems, New Solutions* (New York: Guilford Press, 1998).

2. Andrew Arato, "Luckac's Theory of Reification," *Telos,* 11 (1972); Norman Geras, "Essence and Appearance: Aspects of Fetishism in Marx's Capital," in *Ideology in Social Science,* Robin Blackburn, ed. (London: Penguin, 1972).

3. Rafael Moure-Estes and Theodora Tsongas, "Benzene and Cancer: The OSHA Standard, Workers' Compensation and Public Health Policy," *New Solutions: A Journal of Environmental and Occupational Health Policy* 1(2) (Summer 1990): 13–22.

4. Beth Rosenberg, "Best Laid Bans," unpublished Ph.D. dissertation, Department of Work Environment, University of Massachusetts, Lowell (1997).

5. Charles Noble, *Liberalism at Work: The Rise and Fall of OSHA* (Philadelphia: Temple Univ. Press, 1986); Daniel Berman, *Death on the Job: Occupational Health and Safety Struggles in the United States* (New York: Monthly Review Press, 1978); Peter Donnelly, "The Origins of the Occupational Safety and Health Act of 1970," *Social Problems* 30(1) (October 1982); Nicholas A. Ashford, *Crisis in the Workplace: Occupational Disease and Injury* (Cambridge, Mass.: MIT Press, 1976).

6. Ray Elling, *The Struggle for Worker Health: A Study of Six Industralized Countries* (New York: Baywood, 1986).

7. Benjamin Mintz, *OSHA: History, Law and Politics* (Washington D.C.: Bureau of National Affairs, 1984); Barry S. Levy and David H. Wegman (eds.), *Occupational Health: Recognizing and Preventing Work-Related Diseases* (Boston: Little, Brown, 1995); Michael Silverstein, "Remembering the Past, Acting for the Future," *New Solutions: A Journal of Environmental and Occupational Health Policy* 5(4) (1995):80–85.

8. James Weeks, "Undermining the Protections for Coal Miners," *New Solutions: A Journal of Environmental and Occupational Health Policy* 1(2) (1991): 32–42.

9. Sarah Kuhn and John Wooding, "The Changing Structure Of Work in the US: Part 1—The Impact on Income and Benefits," *New Solutions: A Journal of Environmental and Occupational Health Policy* 4(2) (1994):43–56; Sarah Kuhn and John Wooding, "The Changing Structure of Work in the US: Part II—The Implications for Health and Welfare," *New Solutions: A Journal of Environmental and Occupational Health Policy,* 4(4) (1994):21–27.

10. Charles Levenstein, John Wooding, and Beth Rosenberg, "The Social Context of Occupational Health," in Levy and Wegman, *Occupational Health.*

11. Allard E. Dembe, *Occupation and Disease: How Social Factors Affect the Conception of Work-Related Disorders* (New Haven, Conn.: Yale Univ. Press, 1996).

12. Rita Arditti (ed.), *Science and Liberation* (Toronto: Univ. of Toronto Press, 1980); Sandra Harding, *Whose Science, Whose Knowledge?* (Ithaca, N.Y.: Cornell Univ. Press, 1991); Vincente Navarro, *Crisis, Health, and Medicine* (London: Tavistock, 1986); Vincente Navarro, "Professional Dominance or Proletarianization? Neither," in *The Corporate Transformation of Health Care* (New York: Baywood, 1994).

13. David Swartz, *Culture and Power: The Sociology of Pierre Bourdieu* (Chicago: Univ. of Chicago Press, 1997).

14. Stanley Aronowitz, *Science as Power: Discourse and Ideology in Modern Society* (Minneapolis: Univ. of Minnesota Press, 1988).

15. Noble, *Liberalism at Work.*

16. David Rosner and Gerald Markowitz (eds.), *Dying for Work* (Bloomington: Indiana Univ. Press, 1987); Barbara Ellen Smith, "Black Lung: The Social Production of Disease," in *International Journal of Health Services* 11 (1981):343–359.

17. Antonio Gramsci, *Prison Notebooks* (New York: International Publishers, 1971).

18. Michael Lax, "Workers and Occupational Safety and Health Professionals: Developing the Relationship," *New Solutions: A Journal of Environmental and Occupational Health Policy,* 8(1) (1998):99–116.

4

Brownfields and the Redevelopment of Communities: Linking Health, Economy, and Justice

William Shutkin and Rafael Mares

More than a thousand hazardous waste sites plague Massachusetts' inner-city neighborhoods. The situation is the same in most lower-income neighborhoods and neighborhoods of color in America's central cities. Called "brownfields," these toxic sites not only expose neighborhood residents to environmental and public health hazards, but deter and sometimes prohibit much-needed economic redevelopment. At the same time, in Massachusetts as in most states, the current hazardous waste cleanup laws designed to encourage cleanup of brownfields and deter further pollution have largely failed. This has resulted in endless litigation by potentially responsible parties (PRPs) seeking to avoid liability for contamination that in many instances occurred generations ago.

The brownfields concept, as originally conceived in the early 1990s by policy-makers at the U.S. Environmental Protection Agency (EPA), was intended to solve two pressing social problems with one policy tool. First, by providing incentives to developers and businesses in the form of new, more flexible liability rules and generous funding mechanisms, brownfields programs seek to encourage the cleanup and redevelopment of sites in environmentally and economically burdened communities. Second, as a consequence of the first, such programs are intended to spare clean, more rural or suburban sites—"greenfields"—from development, ameliorating the adverse environmental and social effects of sprawl and destruction of habitat. Environmental and economic benefits are thus intertwined.

The term "brownfields" captures the spectrum of issues that contribute to a community's overall welfare. Decisions about the cleanup and reuse of dirty sites are much more than just land use or environmental decisions. They are decisions about the way in which a community perceives itself and is perceived by others, about the value it places on human health, now and in the future, and the vision it possesses about what kind of place and community it seeks to become. In essence, all brownfield sites tell a story about the history of a community and a place and about people's capacity to rectify past mistakes in creating a better future. Brownfield efforts expose the true breadth and meaning of community health, which is the overall ability of a community to sustain and care for a diverse culture and economy while experiencing a dynamic sense of place, both built and natural.

In this chapter we look beyond the stated purpose of the brownfields concept to explore the full potential of a brownfields program through the revitalization of historically hard-hit communities. At the core of our approach is the notion that the brownfields issue, by nature, demands a long-term, interdisciplinary, and democratic policy perspective, and that brownfields reform should not be the occasion for the continued privatization of the environmental cleanup process. We argue that we must not think about brownfields solely in the legal and technical terms of remediating hazardous waste sites or financing real estate. Rather, we should think about brownfields as an opportunity for economically disadvantaged communities to begin reclaiming their environment and economy and, consequently, for every state to reap the benefits of a cleaner environment and healthier economy.

First we review the brownfields issue and the challenges posed by cleanup and redevelopment of brownfields. Then we detail the opportunities inherent in the brownfields crisis, and suggest specific measures that comprise the foundation of successful brownfields action, including two examples of innovative brownfield redevelopment projects in inner-city Boston. This is followed by a discussion of the importance of planning and innovative environmental strategies for reuse of brownfields and revitalization of communities. We conclude by describing how redevelopment of brownfields is as much about the development of social capital in communities as it is about economic development and environmental protection.

Brownfields Problems and Challenges

Extent of the Brownfields Problem

Over the past several decades, many communities across the United States have experienced a steady loss of industrial and manufacturing businesses owing to suburbanization and white flight, national and global competition, and the recent shift from a manufacturing-based economy to a service-centered one.[1] In addition to countless lost jobs, these businesses have left behind thousands upon thousands of abandoned, contaminated sites, known as brownfields.[2] The EPA defines brownfields as "abandoned, idled, or under-used industrial and commercial facilities where expansion or redevelopment is complicated by real or perceived environmental contamination." Brownfields range from abandoned incinerators to closed gasoline stations to former electroplating facilities.

The U.S. General Accounting Office (GAO) has estimated that between 130,000 and 450,000 contaminated commercial and industrial sites exist around the country.[3] Current estimates place the cost of cleaning up the nation's brownfields at $650 billion.[4] The sites vary in size from less than 1 acre to hundreds of acres. One Boston neighborhood, located near Dudley Square in Roxbury, Massachusetts, covers just 1½ square miles yet contains within its boundaries fifty-four state-identified hazardous waste sites.

Brownfield sites occur all over the country, in urban, suburban, and even some rural communities. Many are located in the Northeast and Midwest where, historically, much of the economic activity was industrial in nature. Both cities and suburbs suffered the effects of industrial flight and economic transformation, but suburbs have generally been able to attract reinvestment capital by expanding other employment sectors. Their inner-city counterparts have been less fortunate. As a result, brownfields tend to be disproportionately concentrated in distressed urban areas, namely, in communities made up of people of color and lower-income groups.

Effects of Brownfields

By nature, brownfields pose a public health risk and an environmental hazard. Until these sites are properly and permanently remediated, contamination can spread to neighboring properties, affecting the soil and groundwater. More important, abandoned sites are frequently left open and unsecured, creating an unattractive nuisance. Even fences and warning signs fail to keep people out. Since they are frequently located in the midst of residential neighborhoods and retail centers, children wander onto these sites and homeless people use them as temporary shelter. Limited only by the level and type of contamination, these abandoned, contaminated sites affect people's health directly and on a daily basis.

The adverse effects of brownfields, however, go far beyond their immediate environmental impact. Vacant lots, contaminated or not, attract illegal dumping activities. The Dudley Square area in Roxbury, for example, with its fifty-four confirmed hazardous waste sites, is also home to over 1,300 parcels of vacant land that is as yet untested for contamination. Many if not most of these vacant lots are used as illegal solid waste disposal sites. These dumping activities bring along their own environmental threats ranging from groundwater pollution to rodent infestations. Furthermore, there is a direct correlation between vacant lots and criminal activity—they are a magnet for all sorts of illicit dealings.[5] Brownfields invariably lower property values in the community, increase insurance rates for neighboring properties, and discourage economic development not only on the contaminated site itself but also in the whole community. The presence of brownfields thus contributes to reduced economic development and creation of jobs.

Barriers to Redevelopment

Despite their deleterious public health, social, and economic effects, most brownfields are not being remediated. Stakeholders present numerous reasons for this phenomenon. While environmental cleanup costs are without question an impediment to the remediation and redevelopment of brownfield sites, according to developers, the most significant barrier is the fear and uncertainty of liability associated with brownfields. For lower-income communities and communities of color where most brownfields are located, however, environmental contamination is only part of a bigger redevelopment problem.

Environmental Cleanup Costs   Data on actual environmental cleanup costs of brownfields are limited. The costs vary from site to site, ranging from a few thousand to millions of dollars. In many cases, the cost can be so substantial as to be prohibitive for economically viable redevelopment. Merely assessing environmental conditions and estimating the anticipated cost of remediation often requires a significant investment. Even when developers pay for a site assessment, they can rarely assume that the estimated cost for remediation is finite. The remediation process frequently uncovers unanticipated areas or levels of contamination. Nevertheless, the actual cleanup costs are rarely the main obstacle to redevelopment.

Fear of Liability as an Obstacle to Redevelopment   Developers cite the fear and uncertainty about liability as the greatest obstacles to redevelopment of brownfields. Owners of abandoned industrial sites prefer to keep these properties idle rather than sell them and take the risk that the environmental assessments required upon sale will detect contamination that they then have to clean up.

In the same way, liability has left investors wary of purchasing properties that may be contaminated and has kept many lenders from financing their purchase or development. Financial institutions now regularly deny credit for such properties, citing environmental concerns. This practice, known as "greenlining," often serves as a legal excuse for continuation of the illegal practice of redlining.[6] After all, most brownfields are located in the same communities targeted by redlining.

Legal liability at brownfield sites stems from frequently overlapping federal and state environmental laws. Depending on the type and extent of contamination, enforcement action may be warranted under the federal Comprehensive Environmental Response, Compensation, and Liability Act (CERCLA or Superfund)[7] and parallel state laws. Many now consider the liability provisions of these laws, passed to ensure the remediation of contaminated sites, as a major impediment to cleanup.

However, in reality, CERCLA reaches only a small number of brownfield sites. CERCLA's National Priority List (NPL) contains only about 1,250 of the nation's hundreds of thousands of contaminated sites. Sites that do not meet the NPL criteria are covered by state legislation, often referred to as mini-Superfund statutes. Mini-Superfunds typically provide the state with authority to force PRPs to cleanup contaminated sites and to establish a state fund to finance state-led cleanups when immediate action is necessary to protect the public health and environment or when no solvent PRPs can be found.

Nonenvironmental Impediments to Redevelopment of Brownfields   In distressed urban areas, contamination and liability concerns are not the only impediments to cleanup and reuse of brownfields. Redevelopment prospects, in general, are bleak in most communities of lower-income groups and people of color. Racial discrimination, segregation, and media-driven misperceptions about crime in inner city neighborhoods are often determining factors in development decisions, ruling out brownfield sites in the minds of developers well before any serious consideration is given to environmental concerns.

Racism and social injustice have created invisible walls around whole communities and have caused decay and prevented revitalization. Siting, zoning, and lending practices, such as redlining, have led to reduced interest in redevelopment. As a result, transportation as well as other infrastructure has deteriorated to the point where it is a significant barrier to redevelopment. Thus, while brownfields exacerbate the problem of redevelopment considerably, overcoming the environmental problems alone will not create a sufficient market demand for most of these properties. It is unlikely that a large number of developers will rush in to develop urban brownfields, even if their liability concerns can be assuaged. As a result, abandoned industrial sites located in suburban areas tend to be more attractive for relief that targets brownfields.

Evolution of State Brownfields Policies

In recent years, most states have begun to address the brownfields problem by enacting voluntary remediation programs.[8] The federal government launched its own brownfields program in early 1995. Powerful financial interests include developers, corporate property owners, the real estate industry, and lenders. On the public interest side, stakeholders include economic development advocates, environmentalists, environmental justice activists, and community members who live next to and are affected by contaminated sites.

Unlike the first-generation environmental laws that rely on enforcement-driven approaches, the brownfields programs count on incentives.

Despite their differences, existing and proposed brownfields programs around the nation are remarkably similar. Almost all share the following components: (1) exemption from liability for causally innocent parties who conduct cleanups, (2) flexible rules that enable parties to voluntarily achieve permanent cleanup standards and get out of the regulatory system as quickly as possible, (3) clear definitions of liability end points for parties willing to tackle brownfield sites, and (4) government-funded financial incentives. There seems to be an unspoken consensus that all that is needed to solve the brownfields problem are economic and legal incentives for private developers.

Another common component of state brownfields reform is to consider future land use in determining the final level of cleanup at a site. In return for the voluntary limitation of future uses and activities on a site, in the form of an activity and use limitation (AUL), less stringent cleanup is permitted. For example, sites intended for commercial or industrial use have a lower remediation standard than those intended for residential development. This has the dangerous effect of limiting the future development of a community by designating a piece of property as industrial in perpetuity.[9] Long-standing and continuing discriminatory practices in land use decision-making, coupled with the legal opportunity to reduce the level of remediation at a site, amount to an explosive combination for most affected communities. Therefore, many environmental justice advocates have been vocal opponents of such AULs, fearing that the future use of brownfield sites will perpetuate past inequities.[10]

So far, federal brownfields policy has not been as ambitious in reducing the liability of PRPs and innocent purchasers as most of the state programs. The federal brownfields policy was spelled out by the EPA in its Brownfields Action Agenda in 1995.[11] This agenda consists of brownfields pilot projects, delisting of sites in the Superfund database, protections for purchasers, memorandums of understanding with states regarding voluntary cleanups, partnerships and outreach, and job training and development programs.

Most environmental justice advocates do not support these federal and state brownfields programs, which further privatize the cleanup of waste sites. After all, it is the market that has for so long neglected inner-city areas regarding environmental cleanup and redevelopment. Possibly, stronger government enforcement rather than a market-based approach would be more effective in managing the forces that caused the contamination. However, by emphasizing the failure of the command-and-control-based federal and state Superfund laws, PRPs, secured lenders, and other private sector interests have succeeded in making brownfields policy almost synonymous with a market-based approach that eases their burden.

On the other hand, no matter how well intended, most government-led environmental protection efforts have also served more white, middle and upper class communities than lower-income ones. For example, there is a considerable difference in the way the federal government has cleaned up toxic waste sites and punished polluters in communities of color than in white communities.[12] Thus, while reliance on the market is not an adequate policy response, the traditional command-and-control and media-based approaches are also inappropriate. Achieving social justice and protecting public health thus require a new policy approach.

Opportunities in the Brownfields Crisis

In spite of the fear that the brownfields issue is a smoke screen for gutting environmental protection, many activists see opportunities and signs of hope in the brownfields crisis. This hope, in part, stems from a wide spectrum of interests who are flocking to the issue, a critical mass that could prove powerful enough to produce positive change. The operative definition of the term "brownfields" could include both the legacy of environmental injustices and the future of environmentally just and sustainable development.

Environmental justice activists have expressed hope that, if done correctly, brownfields initiatives could draw attention to the physical deterioration of the nation's urban areas, allow communities to offer their version of what redevelopment should look like, and permit the application of industrial ecology to a new generation of environmental and land use policies.[13] Environmentalists see a powerful tool for managing growth in brownfields redevelopment, sparing habitat and natural resources from further degradation.

Revitalizing Communities through Brownfields Redevelopment

At present, urban communities are at a disadvantage in competing with suburban areas in attracting and retaining businesses and private sector investment. Economic development experts believe that this competitive gap could be narrowed with the right redevelopment strategy. The missing ingredient in this effort, however, is usually affordable land and buildings. The cleanup of brownfields could contribute considerably to meeting the need for inexpensive property.

The serious environmental and public health problems associated with brownfields provide a strong argument for subsidies or tax breaks to encourage cleanup. With such financial support, the cleanup and redevelopment of brownfields could become feasible for many sites. The resulting investment in land, demolition, and environmental remediation, in turn, would funnel money and employment opportunities into urban redevelopment, not to mention generating much-needed tax revenues.

The subsidies needed for environmental remediation in the inner city would be at least partially offset by the amount that would have to be invested in new infrastructure for redevelopment outside the urban core, since in most cities the necessary infrastructure already exists, however neglected. Moreover, brownfields tend to be near downtown business districts and therefore are ideal for a host of business support services, such as caterers, accounting firms, messenger and maintenance services, and computer repair shops. Sites in the inner city also ordinarily offer a large pool of potential employees.

The cleanup and reuse of brownfields can function as a catalyst for community revitalization. It has the potential to create jobs through remediation and development, attract new businesses, and increase the local tax base. While brownfields are not the only obstacle to revitalization of distressed urban areas, their redevelopment could accelerate the recovery process. Moreover, the efforts to encourage investment in brownfields through planning and marketing might in themselves reduce the reluctance of businesses and lenders to invest in the community because of perceived nonenvironmental factors.

Case Studies in Community Revitalization Through Redevelopment of Brownfields

Brownfields as a Tool for Cultural Preservation: Viet Aid's Effort to Build a Community Center on a Dorchester Brownfield   The Fields Corner section of Dorchester is known as one of the toughest neighborhoods in Boston. Rates of crime, poverty, and unemployment have traditionally been among the highest in the city. In addition, like other inner-city areas, the neighborhood is home to hundreds of brownfields, many of them contaminated by industrial uses long since abandoned and continuing illegal dumping of hazardous wastes such as oils and solvents.

In 1995, Long Nguyen, a young, energetic lawyer-turned-community activist, helped to found Viet Aid (Vietnamese American Initiative for Development), a community development corporation based in Fields Corner that serves the large Vietnamese American population. In the past several years, these Dorchester residents have begun revitalizing the neighborhood, opening restaurants, convenience stores, and other retail establishments. However, persistent linguistic and cultural differences as well as a blighted environment have undermined the residents' sense of place and community, contributing to the recent flight of a significant number of upwardly mobile Vietnamese households from the neighborhood to outlying suburban communities.

In an attempt to integrate the Vietnamese population into the larger Dorchester community and to prevent the further migration of the Vietnamese to the suburbs, Viet Aid decided in 1995 to build a community center to provide job training, English as a second language classes, and housing and counseling services, among other community services. The center's aim is to create a culturally supportive environment for the Vietnamese refugees and immigrants, as well as other local residents, enabling them to participate fully in the life of the community as a whole.

After they selected a site in Fields Corner on which to build the center, Long and his colleagues suspected contamination, given the existing dilapidated structure on the site and its former use as a gas station and paint spraying facility. Teaming up with Alternatives for Community & Environment (ACE), a Boston-based nonprofit environmental law center with experience in brownfields work, Viet Aid developed a strategy for dealing with the suspected contamination.

ACE and an environmental consulting firm prepared reports for Viet Aid on land use and hazardous waste issues associated with the site. They also helped Viet Aid initiate negotiations with the owner of the site, the City of Boston's Department of Neighborhood Development (DND). The city, however, did not cause or contribute to the contamination, which occurred prior to its taking ownership by tax foreclosure. The purpose of the negotiation was to convince the DND to pay for a site assessment in order to arrive at an accurate estimate of the cleanup costs. The DND was eager to transfer title to the property to Viet Aid for a nominal purchase price, not simply to help the project get off the ground, but to try to rid itself of the potential liability for cleanup and third-party damages that any contaminated site poses to past and present owners.

Because Viet Aid is a relatively new, nonprofit organization with little available cash, it could ill afford the risk of taking title to a brownfield, at least not without full knowledge of the extent of contamination and the cost of remediation. By taking ownership, Viet Aid would be potentially liable for cleanup costs and for any damage the contamination might have caused to the property or health of nearby residents. By all accounts, unlike DND, Viet Aid could not assume the risk of the liability.

After two meetings with DND staff in the fall of 1997, Viet Aid persuaded the city to pay for a full assessment of the site's contamination. Later, in January 1998, Viet Aid convinced the city to remediate the site. This success stemmed from Viet Aid's ability to articulate the compelling positive nature of the community center project. For Viet Aid, taking possession of a clean site would allow them to raise funds from pollution-wary lenders while proceeding with construction on a fast-track schedule. For the city, the project's benefits derived from the credit and positive publicity the city could expect to receive from lending its support to environmental cleanup and community development in one of the city's toughest neighborhoods and on behalf of one of its biggest immigrant communities.

Viet Aid will thus take delivery of a clean site upon completion of the cleanup in 1999. However, before taking ownership, ACE and Viet Aid will seek to negotiate an agreement with the state attorney general's office regarding liability protection for any remaining contamination not caused by Viet Aid and third-party lawsuits for damages. The $1.3

million community center project represents a powerful example of brownfields redevelopment as a tool for neighborhood empowerment.

Urban Agriculture in the Dudley Neighborhood: Environmental Cleanup and the Pursuit of Local Food Production   The Dudley area of Roxbury–North Dorchester is one of the poorest neighborhoods in Boston. This diverse community of African American, Latino, Cape Verdean, and white residents has a per capita income of only $7,600, compared with nearly $16,000 for the city as a whole. The unemployment rate is roughly 16 percent. Approximately 32 percent of the Dudley neighborhood's population falls below the poverty level.

Located less than 2 miles from the heart of downtown Boston, the Dudley neighborhood has an extraordinary amount of vacant land, much of it contaminated; almost 21 percent of the neighborhood's land is composed of blighted, abandoned lots, stagnant reminders of the conflagrations of the 1960s and '70s, when property owners set fire to their buildings to collect insurance proceeds after most of their tenants left the community for the suburbs. This legacy of white flight and disinvestment is apparent on almost every block.

In response to the physical and social decay that had afflicted the neighborhood for over three decades, the Dudley Street Neighborhood Initiative (DSNI) was established in 1984 to engage local residents, organizations, and business owners in a community-based planning and organizing effort to create a "vibrant, diverse and high quality neighborhood." The DSNI is the only community-based nonprofit organization in the country that has been granted eminent domain power over abandoned land within its borders. After a planning process in 1987 that produced a comprehensive plan for neighborhood revitalization, DSNI developed community visioning sessions in the mid-1990s. In 1997, the community identified urban agriculture as a key tool in restoring not only a healthy, vibrant local economy but also a clean, high-quality environment.

To that end, DSNI developed the Urban Agriculture Project (UAP), designed to clean up the dozens of brownfield sites in the neighborhood and to redevelop them for food production and value-added food enterprises. The sites will host an assortment of agricultural uses, from greenhouses and bioshelters to gardens, aquaculture, and food processing facilities, each contributing to a neighborhoodwide local economy based largely on food production.

Before this composite vision can become a reality, however, the sites themselves, the feedstock and foundation of the UAP, must be identified and evaluated not only in relation to environmental conditions but also ownership. The DSNI's plan is to establish of pool of sites that can be marketed to developers and users according to the specifications and requirements of the UAP.

A great challenge for DSNI, as for any large-scale brownfield effort, is to gain access to or otherwise generate information about brownfield sites. First, DSNI must find out who owns the sites. If a site is publicly owned, by the City of Boston, for instance, then DSNI can negotiate with the city's public facilities department to gain control of the site or to dispose of it to a third party. This must occur in compliance with the applicable public bidding regulations that govern the sale of public property. For privately owned sites, however, DSNI must identify the owner and the tax status of the property. If the owner has paid all of the property taxes and there are no other liens on the property, DSNI can attempt to convince the owner to sell the property or to develop it as part of the UAP.

Typically, private owners of sites in neighborhoods like Dudley are difficult to contact, let alone to negotiate with. They often live outside the city. Usually they are speculators, waiting for the right opportunity to sell their land and any buildings on it for maximum profit. Only if the owner is delinquent in payment of taxes on the property can the city exercise its power to take the property through foreclosure proceedings. However, because of concerns about liability for nuisance and environmental hazards, in the absence of a compelling redevelopment plan or strong political pressure, public facilities departments are usually averse to foreclosing on properties in inner-city neighborhoods.

Another hurdle is access to the site to conduct an environmental assessment. Without the express permission of the owner or an imminent threat to human health and the environment caused by on-site conditions, neither DSNI nor any other entity can legally gain access to the site. Thus, privately owned sites are very difficult to incorporate in a brownfield redevelopment strategy like that of the UAP.

Even assuming the issue of site ownership can be effectively managed, the problem of environmental assessment remains. Most owners are reluctant to proceed with site assessment work, given their fear of potential legal liability should the assessment uncover reportable or significant contamination levels. The preferred approach is blissful ignorance. Yet, as the Viet Aid example shows, municipal officials can be persuaded to conduct site assessments and cleanups, especially if they understand that the risks of going ahead with an assessment and cleanup are not much greater than the risks associated with owning a contaminated site. Furthermore, the hazardous waste site cleanup laws in Massachusetts already provide certain liability protections for municipal owners who did not cause or contribute to contamination and who attempt to dispose of a brownfield site.

Still in its early stages, the UAP must reckon with problems of ownership and contamination. The best approach appears to be to identify publicly owned sites and, working closely with the city's public facilities department and redevelopment agency, establish an assessment and marketing plan in accordance with the UAP's mission. With the power of eminent domain at its disposal, DSNI itself might consider taking certain properties, but it too must consider the potential liability and the responsibility for maintenance associated with ownership of any neighborhood site.

What makes the UAP so compelling is that it presents a thoughtful, comprehensive plan for community revitalization, on a slightly larger scale than the Viet Aid case. The DSNI has already initiated a meeting with city officials to discuss the UAP, and has solicited financial support from major foundations. In addition, DSNI is about to begin the construction of a bioshelter on a remediated brownfield site across from their offices on Dudley Street. The bioshelter will be a beacon for subsequent UAP activities and will accelerate the food production and revitalization process. Planning is under way to determine what kind of activities the bioshelter will house. More important, the near completion of the project marks the arrival of the UAP into the Dudley neighborhood and of a new urban vision based on our best democratic ideals.

Community-Based and Sustainable Land Use Planning

Historically, discriminatory and environmentally unsound land use decisions have played a powerful role in shaping inner-city communities—physically, socially, and economically. These communities continue to suffer from a nonparticipatory zoning and development process. The brownfields issue offers an opportunity to break this cycle. Since the nexus between past environmental injustices and the brownfields problem is so clear, there is good reason to believe that communities might have a better chance to be heard in land use decisions related to brownfields than in other decisions.

Remediation and reuse of brownfields can have a significant effect on surrounding communities, but this effect will not necessarily be positive. If a brownfields program simply makes it easier for any and all development, the host neighborhoods might well face many undesirable land uses that bring more pollution but few jobs into the community. Depending on the nature of the development, the reuse of abandoned and contaminated property can result in increased traffic and noise, additional pollution, gentrification, few employment opportunities, or a combination of any of these problems. Yet, the right kind of redevelopment could bring to the community some or all of the benefits residents hope for and in the process spark revitalization.

Many residents have a well-developed vision of what their communities could be like. Distressed communities need retail stores and community service centers rather than more waste transfer stations. Community groups, such as DSNI and Viet Aid, have long been advocating land use improvements in the form of urban agriculture, parks, traffic reduction, and other environmental projects.

Community members are inherently qualified to participate in decision-making on brownfields. Environmental protection is a public good requiring public involvement. Besides their interest in being a part of determining their communities' future, their participation brings clear benefits to developers. Through meaningful community involvement developers can secure crucial buy-in for specific projects from local stakeholders at an early stage in the development process. This will reduce the possibility of public opposition through political or legal actions that could, at the very least, slow down the redevelopment process significantly. Moreover, local residents are often a valuable resource in that they have useful site-specific knowledge (about suspected contamination, for instance), information about the community and local political process, and a network of contacts and associations that could facilitate and expedite the development process. A deliberate, democratic approach to community participation allows a developer to make use of this social capital.

However, both the community's right to participate and the developer's interest in promoting public involvement require that community participation occur from the beginning, not as an afterthought. Continuing and well-informed stakeholder involvement is the only way to ensure that the affected community can influence environmental, technical, economic, and land use decisions. Meaningful public participation needs to go beyond the "decide, adapt, and defend mode."[14] It is not sufficient to provide access to information and opportunities to comment. Developers, whether local entrepreneurs or national corporations, must consider the advice offered by the community. Gratuitous public participation is not only an insult to the community but a waste of time.

Pollution Prevention

Brownfield projects can also provide a unique opportunity for application of pollution prevention and industrial ecology principles. The two-sided nature of the brownfields problem—past pollution and future land use—should help to focus decision-makers' attention on the need for pollution prevention. If brownfield sites are cleaned up only to be replaced with a new polluting source, the same problems will reemerge, leaving future generations with another brownfields predicament. In order not to repeat the mistakes of the past, new strategies are essential.

Industrial ecology, a more comprehensive approach to pollution prevention, is nothing more than the idea of sustainable development put into practice. Emphasizing a systems approach to economic development and environmental protection, industrial ecology evaluates the "local, regional, and global materials and energy flows into products, processes, industrial sectors, and economies."[15] It employs mass-flow and life-cycle analysis, among other tools, within and among firms to reveal the total environmental impacts of industrial activity. Industrial ecology projects include ecoindustrial parks and other redevelopment strategies that integrate the concept of industrial ecology into the planning, design, and operation of new facilities.

To date, the brownfields discussion has been focused exclusively on the front end of the problem, cleanup, at the expense of the back end, reuse. Because the discussion has been dominated by developers and lenders, who are principally concerned about potential liability for cleanup and maximizing profit margins and not about other values such as pollution prevention, brownfields policy has focused exclusively on liability issues. There has been little debate about the nature of the development that takes place after a brownfield site is cleaned up. Yet, brownfields policy should logically be about ensuring that redevelopment does not simply recreate a brownfields problem for future generations. It is thus the perfect opportunity to discuss eco-industrial strategies. Brownfield sites are the most compelling reason for embracing industrial ecology; they are also the best candidates for pilot and demonstration ecoindustrial projects.

Unfortunately, redevelopment proposals for inner-city brownfield sites typically involve undesirable uses, such as trash transfer stations, parking lots, or prisons, inspiring opposition from community stakeholders, not support. As a result, the status quo often prevails. Quality ecoindustrial uses have traditionally not been an option for these sites, though they could be.

Redevelopment of Brownfields as a Tool for Managing Growth

Barriers to the development and reuse of brownfields prompt developers to shift their attention to clean sites, so called greenfields, on the urban fringes and in rural communities. To avoid the costs, risks, and complexities of redeveloping brownfields, developers often build on inexpensive, undeveloped land in suburban and rural areas. However, leaving former industrial sites unused makes no sense. The result is not only environmental and economic devastation in the inner city, but also loss of green space and habitat in the surrounding areas, and the sprawl and air pollution that follow.

Therefore, encouraging redevelopment of brownfields furthers the preservation of undeveloped and natural resources while preventing sprawl and pollution. Directing economic growth to areas that are already served by transportation and utilities reduces the environmentally damaging growth that results from the new subdivisions and infrastructure that are part of suburban development.

Conclusion

To succeed in fulfilling the mandate of revitalizing hard-hit communities, brownfields programs must encourage and assist communities in aggressively promoting themselves in defiance of media-driven stereotypes and the ravages of a legacy of disinvestment, white flight, and racism. However, while many inner-city neighborhoods are well positioned to achieve the environmental and economic benefits of a brownfields program, most lack the organized social capital to take full advantage of the opportunities that brownfields present. Most inner-city neighborhoods do not have a network of organizations and institutions designed to handle issues of environmental cleanup and related redevelopment. Community organizations, local businesses, environmental groups, and residents in hard-hit neighborhoods must work with local environment and economic development officials to attract and incubate redevelopment opportunities on brownfield sites and build up a storehouse of experience and knowledge to ensure that over time these neighborhoods will be better equipped to capitalize upon brownfields opportunities.

Brownfields policy is ultimately about much more than legislation or policy. It is a call to arms for economically and environmentally hard-hit communities to challenge severely unwarranted stereotypes and develop their social infrastructure. It is this process, and this process alone, that will help create the healthy, sustainable communities we all want.

Notes

1. John H. Mollenkopf, *The Contested City* (Princeton, N.J.: Princeton Univ. Press, 1983).
2. The term "brownfields" was coined by the Northeast/Midwest Institute, an independent, nonprofit regional policy center. See Charles Bartsch et al., *New Life*

for Old Buildings: Confronting Environmental and Economic Issues to Industrial Reuse (Chicago: Northeast Midwest Institute, 1991).

3. Resources, Community, and Economic Development Division, U.S. Government Accounting Office, "Community Development–Reuse Of Urban Industrial Sites," GAO/RCED-95-172 (Washington, D.C.: 1995).

4. Charles Bartsch and Richard Munson, "Restoring Contaminated Industrial Sites," *Issues in Science and Technology* (Spring 1994):74.

5. Alternatives for Community and Environment, *Beyond Vacant Lots Pilot Project: Dudley Square* (Boston: ACE, July 1997) (unpublished geographic information system study on file with the authors).

6. Redlining is a pattern of discrimination practiced by banks and other lenders who refuse to make mortgage loans, regardless of the applicant's credit record, on properties in lower-income, minority neighborhoods. In the past, lenders actually outlined these areas with a red pencil. Such practice is now illegal. 12 U.S.C. Sec. 2801. Redlining also refers to similar practices by other businesses, such as insurance companies and food delivery services, among others.

7. Comprehensive Environmental Response, Compensation, and Liability Act, 42 U.S.C. Secs. 9601–9675.

8. U.S. Government Accounting Office, *Superfund: State Voluntary Programs Provide Incentives to Encourage Cleanups.* GAO/RCED-97-66 (Washington, D.C.: GAO, 1997, p. 5).

9. Superfund Reauthorization: Hearings Before the House Commerce Committee, 104th Cong., 1st sess. 11 (1995) (statement of Florence Robinson, environmental justice activist).

10. Deeohn Ferris, "Future Use Would Continue Past Inequities," *Environmental Forum* 10(6) (November/December 1993).

11. U.S. EPA, Brownfields Action Agenda (1995).

12. Marcia Coyle and Marianne Lavelle, "Unequal Protection: The Racial Divide in Environmental Law," *National Law Journal* (September 21, 1992).

13. National Environmental Justice Advisory Council, Federal Advisory Committee to the U.S. Environmental Protection Agency, Charles Lee (ed.), *Environmental Justice, Urban Revitalization, and Brownfields: The Search for Authentic Signs of Hope,* EPA 500-R-96-002 (Washington, D.C.: EPA, 1996).

14. Ibid., p. 20.

15. C. Powers and M. Chertow, "Industrial Ecology," in *Thinking Ecologically: The Next Generation of Environmental Policy,* M. Chertow and D. Esty, eds. (New Haven, Conn.: Yale Univ. Press, 1997, p. 24).

5

Place Matters

Mindy Thompson Fullilove and Robert E. Fullilove III

Places are the containers for our lives. The structures of place(s) orders daily existence, making it easier or harder to live our lives and affecting our well-being in important ways. Places serve essential functions in our lives, and are objects of important emotions.

Webster's Ninth New Collegiate Dictionary has approximately twenty definitions (depending on how you count them) for "place," the noun.[1] "Place," the verb, merits another thirteen or so. While perhaps not the most indefinite of words, it is sufficiently vague as to be troubling, particularly when used in a simple sentence such as, "Place matters." "No, it doesn't"—an all too common rejoinder—might be true or might not, depending on our ability to decipher the meaning of "place" in the first place.

From the wealth of possibilities, we focus here on an archaic definition of place offered by Webster's: "the three-dimensional compass of a material object." To investigate this definition, try the following thought experiment:

Imagine a bowl turned upside down. Then put yourself inside the bowl and look around. How do you feel inside the bowl? Now replace the bowl with the following "material objects":

a cathedral

a subway car

a wide meadow under a blue sky

a city street

Each of these material objects creates a different set of feelings within you. Each is a different "place." When geographers speak of the "sense

of place," they are referring to this awareness of that which encom-
passes us.[2]

In each version of the thought experiment it is the interiority to the
material object that we wish to emphasize. "Place" emerges from geography as an essential word for environmentalists because it describes that
bit of the environment within which we find ourselves. The uniqueness
of place is created by the particularity of the material object and of the
moment in which we experience it. If we examine the concept of a landscape, for example, which has come to mean the scenes we observe *from
the outside looking in,* place, by contrast, influences us because *we are
inside* a moment in space and time.

Our work on place has come from a particular vantage point: that
of studying the AIDS epidemic in minority communities in the United
States. Since the beginning of the epidemic, African Americans and
Hispanics have been disproportionately represented among those with
AIDS, and that fraction has steadily increased. By 1996, the proportion of all cases that were black or Hispanic was over 40 percent and the
proportion of minorities among AIDS cases diagnosed that year was even
higher.

Our research has focused on studies directed at understanding this pattern of excess risk.[3] Our most important finding is that in general, risk
behaviors are not higher among these populations. Rather, we learned
that they are focused in a small number of sociogeographic communities.
It was through the work of Dr. Rodrick Wallace that we began to understand the peculiar contribution of location to the spread of infection.[4]
Specifically, his ecological studies underscored the observation that widespread destruction of housing in particular minority neighborhoods defined the areas most at risk for high rates of HIV infection. As we began
to focus our studies on those areas—the South Bronx and Harlem in
particular—we found that numerous other conditions were colocated in
the same physically devastated areas. The role of high risk settings—what
we began to call "toxic environments"—in the spread of disease became
the focus of our work.[5] In particular we needed to know why drug use
was so prevalent in these physically devastated areas.

To answer that question, we looked in great detail at the environments
themselves.[6] Between 1992 and 1994, we conducted a population-based

survey of over 700 residents of central Harlem, using stratified random sampling techniques.[7] In addition to talking to the residents about their health status, we examined the physical condition of the forty randomly selected blocks on which our respondents lived.

The picture at every level was grim. Housing in the area was in poor condition, and many buildings had been abandoned. A great number of blocks had lost one or more buildings; vacant lots, often strewn with garbage, were everywhere. In the survey area, it was difficult to find a two-block stretch that did not go past a vacant lot or abandoned building.

Those of us involved in data collection spent many hours "inside" the streets of Harlem, ingesting the sights, sounds, and smells of a deeply injured urban habitat. It was not hard to imagine the links between the degraded physical conditions and rates of substance abuse in these neighborhoods that according to our survey were two to five times the national rates for African Americans. This is not to suggest an overly simplistic determinism between place and health, but rather to underscore how oppressive it is to be within a disintegrating place. Instead of having one's life simplified or uplifted by one's surroundings, everything is made heavier and more complicated.

The degraded environment of Harlem is toxic to the health of residents, as measured by rates of asthma, violence, addiction, AIDS, or a host of other diseases. Yet a few blocks away—on the upper East Side—the relative health of the population is much better. Wallace's ecological studies demonstrate that health is shaped within small areas. As the geographers would describe it, health is shaped by the environment within a particular place. There is a very literal foundation for the idea "Think globally, act locally," which is that the immediate environment has an extraordinary effect on physical and psychological health.

The person-place relationship is a partnership. In order to understand this relationship, it is essential to understand the ways in which people relate to their places. It is also essential to be aware of the services that places provide for people. In the next section we describe the two sides of the person-place relationship: first the "person" side and then the "place" side. With those ideas in mind, we turn to health care settings to illustrate some of the principles we have put forward. Then we return to the problem with which we started, AIDS in Harlem, to focus on the ways in which the place has failed the people. At that point we will be able to explore the ways in which a new vision might lead us to a healthier future.

The "Person" Side: Belonging to Place

A large body of work in many disciplines, including social psychiatry, geography, anthropology, and environmental psychology, offers insights into the relationship between person and place. Our synthesis of this literature, complemented by case studies, suggests that people form profound relationships with their intimate environment. In order to understand these relationships, we have proposed a theory of the "psychology of place."[8] This theory is focused on three elements in the person-place relationship: the sense of belonging to place, the individual's need for a dense matrix of social interactions, and the awareness of the interdependence of places.

The sense of belonging arises from a secure sense of being a part of a place. This depends on the accomplishment of three psychological tasks. People internalize a map of the spatial organization of any given place, a process called "orientation." People connect to a place because it provides them with essential life supports, such as shelter from the elements and protection from predators; that is, they develop attachment to place. People derive a portion of their identity from the places where they live and work, forming a "place identification." The successful accomplishment of these three tasks provides the basis for taking comfort and security from a place.

The sense of belonging is augmented by the establishment of a satisfying routine of interactions with other people who live and work in the same location. In places that are "good enough" to support health (an idea explored in depth later), individuals interact with large numbers of other community members. Such a dense matrix of social connections facilitates many community functions, such as the communication of group norms. The matrix of interactions is formed by the existence of places within the place: schools, shops, community centers, work sites, streets, parks, churches, union halls, swimming pools, and others.

A healthy sense of place must balance the needs of "my place" with the needs of "their places."[9] Places are nodes within a complex environment and are dependent on each other for the sustenance of the common good. This reality should be the basis for cooperation and sharing. However, quite the opposite is true. The existence of well-tended places side by side with neglected places, just as the wealthy upper East Side sits cheek-by-jowl with the impoverished community of Harlem, teaches us that an ethos of competition dominates relationships among communities.

From the psychology of place, we can infer a great deal, not only about what people feel about places, but also about what people need to receive from places. Place, as an object of love, is not the passive recipient of our projections, but an important actor shaping our lives. What is it that people need from place?

The Place Side: Being "Good Enough" for People

We approach this issue by following the lead of psychiatrist Donald Winnicot, who observed that infants do not need *perfect* mothers, but they do need a minimal level of nurturance, attention, and love. Mothers have to be, in his words, "good enough." Similarly, we argue, places must fulfill basic sustenance needs for human beings and therefore must be "good enough" to support life. What does this entail?

A "good enough" place *has a center and a boundary.* A good enough place is a place where people want to be. They are drawn there by a kind of magnetic force that pulls them toward the place. The places with the strongest power of attraction have a central focus that pulls people in. For example, a well-designed neighborhood will have a town square that draws people to the center. Conversely, at the end of the place, there is a boundary. The boundary is essential because it defines in and out, giving the place its essential interiority. Almost everything about human interaction is based on the tension between interior and exterior.

A "good enough" place makes sense to the people within it. Places are organized according to culturally derived rules. Assuming that places follow the rules, people within the place know how to use what is there.

However, sometimes the rules are violated, and this can be distressing and disorienting. It may help to describe a typical block in Harlem. Manhattan, like some other sections of New York, is built to resemble the housing equivalent of a canyon, with tall, tightly juxtaposed buildings rising to flank the sides of narrow streets. This canyon architecture used to cover all of Harlem, typically with buildings between six and ten stories in height. Between 1960 and 1990, approximately 30 percent of the housing was destroyed, leaving holes in the formal canyon structure and destroying the visual coherence of the form. Whereas what one would expect to see in an urban canyon is building, building, building, etc., all the way around a block, what one now sees in Harlem is building, building, abandoned building, vacant lot, etc.

A "good enough" place *provides physical sustenance.* People obviously require all sorts of environmental sustenance. They need food, sunshine, exercise, clean air, shelter from the elements, and protection from predators. People must also be able to rid the environment of toxic substances, such as wastes and other poisons.

A "good enough" place *provides existential support.* Life, in all its incomprehensibility, is stressful for human beings. It is essential that they have support in managing the incomprehensible. Thus, religious institutions, cultural events, and other activities are essential to managing the joy, wonder, suffering, and other difficult parts of the human condition.

A "good enough" place *is linked to other places.* For many reasons, not least of which is curiosity and a wish for variety, people cannot be confined to a single place. Rather, they need to be able to travel, visit, immigrate, explore, pioneer, vacation, and otherwise investigate the wonders of the earth. In order for this to happen, places need to be linked to each other. In earlier centuries, places were sufficiently isolated that it was possible for one place to exist independent of what happened elsewhere. This is no longer the case. The corollary of places being linked is that places are mutually interdependent. A good enough place respects the rights and needs of other places.

In sum, good enough places work for the people within and do not harm the people without. Understanding how places work for people is an idea we will explore further by looking at health care settings.

Health Care Settings as "Good Enough" Places

The Picker-Commonwealth Program for Patient-Centered Care sponsored a large study of patients and their families designed to find out what people wanted in hospital-based care and if they felt they were getting it. In the report on the study, *Through the Patient's Eyes*,[10] Gerteis and colleagues introduce the issues of concern to them with the following story:

> Some years ago when officials in Massachusetts tried to close a small and foundering community hospital in the north-central part of the state, busloads of loyal patients and angry citizens descended on the State House in protest and ultimately fought the effort all the way to the Supreme Judicial Court of the Commonwealth. . . . There were good and well-known reasons for closing the hospital. . . . Why, then, all the hue and cry? . . . Townspeople preferred to be sick in their own hospital. There, you were among friends and neighbors who most likely knew you, your family and your situation without having to be told. . . . "I had occasion to be at your Mass. General Hospital when my brother was operated on," [one woman] wrote in a letter to the governor. "As to the care he got well, I would put our facility up against yours anytime. No bath or back rub for a week and a room that looked like a prison cell. Our cheery little hospital would put Mass. General to shame."[11]

Gerteis and her team concluded that planners had failed to appreciate that the patients' perspective on caregiving may be quite different from that of other actors in the health care industry. Their study was able to uncover a range of needs patients would like to have met: maintaining a sense of individuality, getting information, integrating family and friends into the treatment process, being comfortable while in the hospital, and support while making the transition back to home. Some hospitals, Gerteis found, often failed across the board at creating a "patient-centered environment." Those that succeeded often did so because of a deep, philosophical commitment to making patients the focus of the institution, an idea that became a reality because it was embedded in the physical and social structures of these places.

Health care settings, like other settings, order behavior in large part through the cultural expectations attached to the ordering of space. Psychiatry offers many examples of this principle. Sigmund Freud, the founder of psychoanalysis, believed that the analyst must be present as a blank screen for the analysand's projections, which would be revealed by free association, dreams, and slips of the tongue. The treatment environment was conceived as a couch placed at an angle to the analyst's chair. While the analysand relaxed and engaged in free association, the analyst sat just out of sight and guided the process. The analyst's couch is so famous that it is a staple of movies and cartoons. In fact, in the day-to-day practice of the profession, to walk into a colleague's office and see a couch leads automatically to assumptions about the nature of their practice.

Child analysts, by contrast, recognized that other environments were needed to access the secrets of the developing mind. Play corners and doll houses are standard equipment for those who treat children. A fascinating subset of child analysts consists of the Jungian analysts, who use sandboxes accompanied by remarkable collections, sometimes numbering in the thousands, of small objects with which to create imaginary worlds. The display of these objects so that they are easily accessible to children is always incorporated into the design of the Jungian treatment space.

Day treatment centers provide another example. Because the goal of day treatment is to encourage socialization, there are always large rooms in which the treatment community, composed of staff and patients, can gather for meetings or other social activities. The preparation and serving of food is a common activity, both because it promotes minimal levels of functioning and because it supports the interactions of people within the group. Medium-sized rooms for groups and small rooms for offices are usually included in the plan, but at least for day hospitals, are secondary to the necessity for a large, open space.

Roger Barker, who developed the concept of ecological psychology, noted that there is a usually a conformance between the organization of a space and the behavior that will go on there.[12] A church, with its pews facing the altar, supports a worship service, but is not a good place for a dance. A baseball field is perfect for baseball, but not for football (just ask the San Francisco 49ers!). A mismatch between space and activity—trying to conduct psychoanalysis in a gym or playing baseball in the living room, for example, can seriously undermine either the space or the activity or both. Part of the process of growing up is learning what should be done where. Barker found that in fact the rules of "behavior settings" explain much of what we see people do.

Behavior Settings

In Barker's theory, a behavior setting is a natural phenomenon like weather or topography. Such a setting is a bounded event that occupies a designated space at an identified time. It is composed of people and things. There is a "synmorphic" relationship between the nature of the activity and the arrangement of necessary objects. The behavior that goes on in the setting is independent of the actors there at any given time. However, neither setting nor behavior can be greatly altered without changing the nature of the behavioral setting. Barker, in a survey of a Midwest town, identified 220 distinct behavioral settings, including: 1 animal feed store; 2 automobile washing services; 3 award ceremonies; 25 bowling games; 2 bus stops; 8 card parties; 1 cemetery; 1 drugstore; 13 elementary school basic classes; etc.

An example may provide further clarification of this essential idea. Little League baseball games are behavior settings. Parents, coaches, and guests show up to watch little boys and girls execute the program "baseball." This program requires two teams of nine people who play for a certain number of innings and follow a well-known set of rules. Little League games vary in the extent to which the adults view themselves as preparing future Big Leaguers or promoting socialization, skill development, and fun. Is competition an essential part of "baseball"? Should it be part of Little League baseball? These are questions that people ask constantly as they seek to adjust the functioning of behavior settings without losing their essential components.

Feelings about Settings

Cultural change and social reorganization often force us to examine the meaning of the settings and the behaviors in our lives. The hospital closing story cited earlier is an example of a threat to a setting because it no longer met professional and technical standards for a hospital. Local residents were distressed, and rightly so, because as a behavioral setting, their local hospital offered gracious, personal care. The new hospitals, by contrast, had better equipment but a less personal touch. Such an alteration in the environment can create a sense of alienation in a patient who feels that his or her need for personal care no longer matters.

How do we organize behavioral settings that meet the needs of a changing and more varied group of users? A Little League parent who wants his son to have fun, a patient who wants a bath: These are people expressing their feelings about the ways in which they would like behavioral settings to operate.

The feelings of people are an essential element in understanding the dynamics of behavior settings. Behavior settings exercise a remarkable degree of control over our actions. Our conformity does not necessarily indicate contentment. To go beyond the behavioral compliance—to understand if people are satisfied—it is essential to ask their opinions in free-wheeling, open-ended interviews. To understand how people feel about the places in their lives, it is important to listen for orientation, place attachment, and place identity as key threads in the narrative.

Making a Behavior Setting Work

The investigation of childhood sexual abuse is one of the most difficult areas in modern medicine, yet appropriate interventions are essential because of the severe implications for growth and development and adult functioning. At most hospitals in New York City, an abused child is brought to the hospital emergency room, where portions of the investigation are conducted. Later, if there is evidence to justify police involvement, the child may be taken to the police station for furthering questioning. Neither emergency rooms nor police stations are child-friendly environments. In a further complication, the involvement of many different agencies is needed in order to process a case. The agencies are scattered and often understaffed. Getting people together to discuss a case or to conduct an investigation simultaneously is considered ideal by all involved in this work, but it is nearly impossible to achieve.

A breakthrough in this inefficient and distressing state of affairs occurred in 1997 when Victim Services and the City of New York jointly established the Child Advocacy Center. The center, located on the East

River in Brooklyn, includes two wings. One wing is bright and sunny and is dominated by a beautiful mural of sea creatures and a large playroom. The furniture is built to child height and the spaces are cozy. The other wing looks like a standard office, with dozens of big metal desks and file cabinets crammed together, phones jangling, and copying machines humming. There, police, child welfare workers, and health care practitioners can jointly conduct the work of investigating allegations of abuse. In this extraordinary dual space, the abused child is centered in the child wing, while adults move back and forth between the two spaces.

The creation of such a space has resulted in remarkable alterations in the behavior of all involved. The children are much more comfortable and more likely to tell their stories. The adults, gathered as they are from disparate bureaucracies, can achieve a shared understanding of a case because they can conduct the interviews together. The Child Advocacy Center expected to increase arrest and conviction rates and improve the treatment of children. The early results suggest that it is meeting those goals. The center has also reported that it is overwhelmed with business, which was not expected. This response probably reflects the logic of the behavior setting, which is closely aligned with what children need and adults want. The relief and excitement expressed by normally stoic New York City detectives and welfare workers is perhaps the most remarkable testimony to the efficiencies to be gained by having the "right" space.

From Health Care Settings to Communities

We have applied the general principle of "good enough" place to the organization of health care settings. From the brief discussion given here, we can see that health care organizations must first recognize the needs of various and perhaps competing groups of participants, and then organize the spatial relationships so that the groups can work together. These two issues are the core issues in the organization of all places. Communities are larger than health care facilities, but share with them the problem of bringing many competing interest groups into cooperative relationships with each other. In addition to harmonizing social groups, communities must also create links among the many settings that are its constituent parts. Thus, for a community to be a good enough place, it must be composed of the right parts, and the parts must be well synchronized.

In our earlier descriptions of Harlem, we suggested the devastation of the landscape. We can return to that theme from the perspective of enumerating the parts and assessing their synchronization. Two features emerge from our research. First, what had been a rich matrix of behavioral settings has been impoverished by the loss of buildings and people. While Harlem used to have hundreds of community organizations, many have disappeared, Little League baseball being but one example of the contraction of the social matrix. Second, the collapse of the landscape made it more frightening and less attractive for people to move about in the community. Some blocks were taken over by illegal activities; "honest" people learned to stay away from those areas. The disintegration of the Harlem community involved both a loss of essential parts and the deterioration of the connections among the units. The once dense matrix of behavioral settings became quite frayed. The decline in health is, in our view, an inevitable outcome of the collapse of place.

However, the collapse of Harlem cannot be understood simply from within the community. Rather, it is essential to view the impoverishment of this area in relation to the enrichment of other areas. Unequal distribution of resources among communities means that some communities, such as central Harlem, are deprived of a fair share of civic services, such as street cleaning, sanitation, building inspection, etc.; of societal investment; and of opportunities to make connections with other places. These deprivations have led to the economic, social, and physical collapse of those places.

The process of favoring wealthy communities has resulted in a bonanza of land cleared by civic neglect. New policies are now being inaugurated to take over land-rich communities and transform them into homes for the well-to-do. In Harlem, this process of gentrification means that the poor, black people who currently live there will be dispersed to other communities. The present-day sociogeographic community of Harlem will be divided into a social community of poor people who are former Harlemites and a geographic community of well-to-do newcomers.[13]

This displacement of the poor is a worldwide phenomenon that has serious consequences for many populations. It may be inferred from the psychology of place that people need the stability and identification that is offered by a home place. When that home place is taken away, people are thrown into confusion and mourning from which they may or may not recover. What is essential, from the perspective of place, is to understand the extent to which the current displacement of populations is a global process.

In the modern global economy, places are only nominally separate from one another. Increasing global interaction and interdependence have transformed the process of community development. No longer are civilizations acting independently. In the modern world, the garbage disposal decisions of one group can radically alter the health of another group, particularly if their community becomes the dumping ground. These decisions, we would posit, are a fundamentally spatial rather than racial process, in which goods are concentrated at the sites that are powerful and wastes are discarded at sites that are weak.

The spatial processes of racism follow this fundamental gradient, concentrating "good people" (mostly white) in the power spots, and "waste people" (mostly people of color) elsewhere. Such efforts to concentrate goods and discard wastes may be thought of as a process of "stripping and dumping," which has transformed the world from one in which many independent communities were each able to create health, into a world dominated by an intercommunity battle for the fundamental goods that make health possible. Although it is generally accepted that it is possible to "win" at this battle, such an assessment ignores the deeper ecological connections that link all communities on the globe.

A fundamental tenet of the concept of "good enough" place is that a good enough place does not harm other places. However, such a concept is not part of people's everyday thinking about the organization of space. People are content to rate a single place according to its qualities, but in an interdependent world, all places must be rated by the extent to which they help or injure other places.

Conclusion

The creation of good enough places depends on global sharing and global cooperation, not the creation of have and have not peoples and places.

It is important to understand the threat to ecology and health of this larger process of stripping and dumping.

If we accept that "place matters," then we must accept the admonition to "Think globally, act locally." Returning again to AIDS, which brought us to concerns with place, this worldwide pandemic is a problem for vulnerable communities all over the globe. In order to control the epidemic, politicians, community activists, and citizens must be willing to tend the physical and social infrastructures of damaged places, whether the damage is from war, civic neglect, or economic displacement. Jonathan Mann, Daniel Tarantola, and Thomas W. Netter, editors of *AIDS in the World: A Global Report,* provided a wealth of data to demonstrate what Wallace had argued in the U.S. context: that excess risk for AIDS is closely linked to the vulnerability of local communities.[14] They reminded us all, "As individuals we are linked to the whole; how to connect the individual and the global, the personal and universal—this is the destiny and special challenge of our time."[15]

Notes

1. *Webster's Ninth New Collegiate Dictionary* (New York: Merriam-Webster, 1990).

2. The contrasts among places are commonly evoked to establish the ways in which place affects the person. Gaston Bachelard, *The Poetics of Space* (Boston: Beacon Press, 1994) is the classic effort of this sort. *The Experience of Place,* by Tony Hiss (New York: Knopf, 1990) and *The Power of Place* by Winifred Gallagher (New York: HarperPerennial, 1994) are more recent important contributions to our understanding of the "three-dimensional compass of a material object."

3. We began our work in AIDS with Multicultural Inquiry and Research on AIDS (MIRA), a component of the UCSF Center for AIDS Prevention Studies. In 1990 we joined the faculty of Columbia University. The Community Research Group was established in 1992 to extend our work on the community issues underlying the spread of AIDS.

4. Rodrick Wallace, "A Synergism of Plagues: 'Planned Shrinkage,' Contagious Housing Destruction, and AIDS in the Bronx," *Environmental Research,* 47 (1988):1–33.

5. Mindy Thompson Fullilove, "Environments as Toxins: Comment on Panel on Strengths and Potentials of Adolescence," *Bulletin of the New York Academy of Medicine* 67 (1991):571–573.

6. The data reported here on Harlem are taken from a series of research efforts: the Harlem Household Study, a survey of a multistage probability sample of Harlem residents, 1992–1994; the Coming Home Project, an ethnographic study of housing resettlement, 1994–1997; and Elder Stories, a collection of oral histories of long-term residents, 1997–1999.

7. Robert E. Fullilove, Mindy T. Fullilove, Mary E. Northridge, Michael L. Ganz, Mary T. Bassett, Diane E. McLean et al., "Risk Factors for Excess Mortality in Harlem: Findings from the Harlem Household Study," *American Journal of Preventive Medicine* **16**(3S) (1999):22–28.

8. Mindy Thompson Fullilove, "Psychiatric Implications of Displacement: Contributions from the Psychology of Place," *American Journal of Psychiatry* 153 (1996):1516–1523.

9. Edward Relph, in the chapter "Sense of Place," in *10 Geographic Ideas that Changed the World,* Susan Hanson, ed. (New Brunswick, N.J.: Rutgers Univ. Press, 1997, p. 222), calls the overvaluing of one's own location a "poisoned sense of place."

10. Margaret Gerteis, Susan Edgman-Levitan, Jennifer Daley, and Thomas L. Debanco, Introduction, in *Through the Patient's Eyes: Understanding and Promoting Patient-Centered Care,* Margaret Gerteis, Susan Edgman-Levitan, Jennifer Daley, and Thomas L. Debanco, eds. (San Francisco: Jossey-Bass, 1995, pp. 1–15).

11. Ibid., pp. 1–2.

12. Roger G. Barker, *Ecological Psychology: Concepts and Methods for Studying the Environment of Human Behavior* (Stanford, Calif.: Stanford Univ. Press, 1968).

13. Lisa W. Foderano, "For Affluent Blacks, Harlem's Pull Is Strong," *New York Times,* January 18, 1998, pp. A1, A25.

14. Jonathan Mann, Daniel J. M. Tarantola, and Thomas W. Netter (eds.), *AIDS in the World: A Global Report* (Cambridge, Mass.: Harvard Univ. Press, 1992).

15. Ibid., p. 843.

6

Akwesasne: A Native American Community's Resistance to Cultural and Environmental Damage

Alice Tarbell and Mary Arquette

The term "hazardous waste" has many meanings to many different people. For those people living in the Mohawk territory of Akwesasne, dealing with hazardous waste issues has become a daily challenge. Toxic substances such as polychlorinated biphenyls (PCBs), dioxins, dibenzofurans, volatile organic compounds (VOCs), polycyclic aromatic hydrocarbons (PAHs), fluorides, cyanide, aluminum, arsenic, chromium, copper, and styrene (to name just a few) have polluted area ecosystems, affecting both the natural world and the Mohawk people. While it is critical for the Mohawks of Akwesasne to tell their own story about the effects of pollution on their territory, people, and resources, there are many difficulties. For example, how can we discuss the effects of toxic substances with elders? For them the mere thought that someone would intentionally disrupt the natural world and interfere with the cycles of creation is such a foreign concept that there is not even a word for pollution in the Mohawk language. How can we work effectively on toxicant issues when so little research has been conducted involving native peoples and so few culturally appropriate tools exist? How can we discuss issues involving cultural, human health, and ecosystem impacts with federal and state agencies when their narrowly focused risk assessment models are often completely inappropriate for native nations, who long ago developed integrated, holistic models of health?

In this chapter, we address the implications of the disproportionate effects of pollution on a Native American community—the Mohawk people of Akwesasne—and the need for federal, state, and provincial governments to respect the cultures of peoples who have a completely different world view from mainstream society. We will show that Akwesasne is a community resisting pollution's damage to human health, cultural diversity, and the natural world.

To date, many confrontations, especially with native peoples, have occurred as a result of environmental issues. In some respect, these confrontations have arisen because cultures have a different stake in the siting, regulation, or remediation of environmentally hazardous activities. When two cultures clash on determining the future, who gets to define what is a rational course of action? Whose values and belief system are more relevant, more respected, more "real"? Mohawks, like many native nations, have a very different world view of their land and resources and this view must be respected in order to maintain cultural and biological diversity, not just in the United States and Canada, but worldwide.

Background

The St. Lawrence river valley has been an integral part of the homeland of Mohawk people for millennia. The St. Lawrence River, known as "Kaniataraowaneneh," and its tributaries have supplied a constant source of fresh, clean water and support countless species of plants, birds, and wildlife. The oldest known settlements of native peoples in the region are located on islands in the river, some of which date back to the Archaic period, more than 9000 years ago.[1] The territory of Akwesasne, located along the St. Lawrence River between northern New York and western Quebec–eastern Ontario serves as home to more than 12,000 Mohawk people who continue to depend on the abundant natural resources in the area. The Mohawks call this region Akwesasne, which means "land where the partridge drums."

This name, like many other words, is an excellent example of how the natural world is an integral part of Mohawk language and culture. Akwesasne includes the land and waters where the Mohawk people have raised their families, fished, hunted, and buried their dead for thousands of years. Oral history tells us that the St. Lawrence river valley has always been a favorite hunting and fishing location, owing to its abundant natural resources and rich, fertile soil.

Over the course of millennia, Mohawk people have developed a unique relationship with their environment. Many social, cultural, and spiritual values, beliefs, and philosophies are tied to their relationship with the natural world. These ties distinguish Mohawks as a unique, culturally distinct people. For the Mohawks of Akwesasne, cultural practices include all areas in their aboriginal territory that have been and continue to be used by Mohawk people. Language, ceremonies, and other cultural activities surrounding fishing, hunting, trapping, gathering, and agricultural activities have always kept the Mohawk people in daily contact with all parts of the natural world. They require a healthy natural world for their physical, mental, and spiritual health.

Almost a thousand years ago, the Mohawk people were the first among many nations to accept a message of peace, known as the Kaianarekowa or the Great Law. As a result, they became part of a confederacy of five nations, known as the Haudenosaunee (or Iroquois). This confederacy was one of the first United Nations in the world based on principles of peace and cooperation. It not only extended these principles to other human beings, but to the natural world as well.[2] The foundation of the longhouses of the five nations rests firmly on our mother, the earth. Because of this, traditional leaders must always consider three things when making any law: the effect of their decisions on peace; the effect on the natural world; and the effect on seven generations in the future.[3] This Great Peace includes the entire creation, and the Haudenosaunee have worked hard to ensure acknowledgment of the voices of the natural world. They have understood that all peoples, including plants, animals, and the earth herself must be included in defining environmental justice. As a result, the Haudenosaunee have been recognized as leaders in the area of environmental protection and restoration.[4]

Many treaties were signed between the Haudenosaunee and European and U.S. governments. Very early on, the Haudenosaunee realized that non-native people had a world view very different from their own. One of the first treaties between Europeans and the Haudenosaunee, called "Kaswenta," recognized these differences. It described the two cultures as two separate vessels in the large river of life. The European and U.S. governments were located in a ship and the Haudenosaunee in a canoe. Each vessel included all the people, lands, resources, laws, government, beliefs, values, customs, and world view of each culture. The vessels were separate entities that were not to interfere with each other. Although separated by the river of life, the signers of the treaty realized that from time to time there would be a need to work together on common issues, and they agreed to use principles of friendship, respect, and cooperation. To this day, the Haudenosaunee have always honored this treaty. On the other hand, U.S. and Canadian governments have often interfered, not only with the Haudenosaunee, but also with the river of life.

History of Environmental Contamination

Time changes many things, but not all changes are for the better. What some people call progress, others call destruction. In the last two generations, Mohawks have seen many negative environmental changes in their homeland, none more devastating than the industrialization of the St. Lawrence River. The master plan for this development included the construction of the St. Lawrence Seaway, the building of the Moses-Saunders power dam, and the industrialization of the region. The "need" to "develop" the river was expressed by project developer Robert Moses.

Moses's ideas for economic and energy development were the keys to his master plan and were to have the most impact on Mohawk life along the St. Lawrence River. . . . By developing public hydroelectric power along the St. Lawrence . . . he would stimulate heavy industry and . . . seaway transport. By constructing . . . parks and parkways for tourism and recreational purposes, while providing special low rates for Saint Lawrence residents, he would counter any local opposition to the project. By improving the state's total economic picture, he would satisfy the utility companies' quest for increased profit margins. By sacrificing Indian lands or those that were claimed by Indians, who were [a] small powerless racial minority largely outside of the American electoral process, he would not alienate white voters and their political representatives. . . . Dams, reservoirs, and power development were part of the 1950s idea of progress and were seen as more important than Indians and the protection of their treaty rights.[5]

The Seaway was constructed in 1954 to open markets and ports in the Great Lakes basin and encourage industrial development in northern New York State. In 1957 the Moses-Saunders power dam brought cheap hydroelectric power to everyone in the region but the Mohawks. As part of the package deal, several industries moved to the region, attracted by the promise of cheap hydroelectric power provided by the dam. The General Motors (GM) Corporation-Central Foundry Division, Reynolds

Metals, and the Aluminum Company of America (ALCOA) now use the inexpensive electricity to operate facilities upstream from Akwesasne.

These industries have contaminated the St. Lawrence River and its tributaries, creating one of the largest sites in the United States contaminated by polychlorinated biphenyls, polyaromatic hydrocarbons, phenols, volatile organics, fluorides, and metals. Toxicants released from these companies have contaminated Mohawk lands, air, and waterways, endangering traditional land use and subsistence fishing.

Through the dam, the New York Power Authority, a public utility that owns the dam, subsidizes the environmental contamination of the St. Lawrence, Raquette, and Grasse rivers and their ecosystems. Many of the toxicants found in the rivers are so persistent that they will affect the Mohawk people for generations.

"Cleanup" of Area Superfund Sites

Many studies have identified the industrial facilities operated by GM, Reynolds, and ALCOA as sources of toxic substances, such as PCBs. As a result of past illegal disposal and storage practices, General Motors was fined $507,000 and their Massena facility was placed on the Superfund National Priorities List in 1983. The Reynolds Metals and ALCOA facilities are state Superfund sites. To be given this status, sites must be identified as a danger to public health and warrant the highest priority in developing and implementing cleanup plans. Today, all three Superfund sites are at various stages in the remediation process.

During recent years, both industry and policy-makers have created increasing pressure to relax cleanup standards in order to reduce costs at Superfund sites, including those located adjacent to Akwesasne. For example, remediation of the heavily contaminated industrial landfill at the General Motors Superfund site cannot even be considered a cleanup. The 12-acre landfill (better referred to as a dump, since it is unlined and cannot be accurately defined as a landfill) is located within 500 yards of homes and businesses in Akwesasne, and is situated on the banks of the St. Lawrence River near an important wetland area. Mohawks have argued for the excavation and removal of contaminated material in the landfill, but they have largely been ignored.[6] At the GM site, the U.S. Environmental Protection Agency (EPA) has chosen a remedy that includes capping contaminated materials in place rather than the more costly option of permanently treating the materials. Mohawks have complained that this is not a cleanup but a coverup and certainly not a permanent solution.[7] The EPA, on the other hand, concluded that leaving the waste on site would be the cheaper alternative since excavation and treatment would double the cost of the remedy. In many respects, the decisions made by EPA to place cost above protection of Mohawk culture, future generations, and the natural world display the difference in how decisions are made by non-native government agencies.

Current policy and law mandate that federal agencies consider the special legal status of native nations. The EPA, like other federal agencies, has a statutory obligation to act as a trustee of the land and resources of native nations. In addition, in 1984, EPA's Policy for the Administration of Environmental Programs on Indian Reservations (which was affirmed in 1994 by EPA Administrator Carol Browner) recognized tribal governments as sovereign entities and agreed to work directly with them on a "government to government" basis.

On Earth Day 1993, President Clinton issued Executive Order 12898, in which he mandated that every federal agency make the achievement of environmental justice part of its mission. The executive order prohibits discrimination against poor communities and communities of color in federally funded environmental programs, including state programs that receive federal funds and those for which EPA has oversight.[8]

In spite of the U.S. federal trust responsibility, the environmental justice executive order, the Indian policy, and the internationally acknowledged importance of native peoples by the United Nations, environmental protection for Mohawk lands and culture continues to erode. Relaxing treatment standards and promoting substandard, temporary cleanups at Superfund sites runs completely counter to the mandate of the executive order on environmental justice. Because the United States has not respected the decisions of the governments like those at Akwesasne, it continues to fail in its responsibility to protect native communities from damage inflicted by U.S. citizens.

At Akwesasne, tough decisions about remediation at area Superfund sites are often biased toward industry and the local economy at the expense of native peoples and their endangered cultures. The Mohawk people and their governments make decisions based on different criteria. Mohawk tradition demands that we accept responsibility for our actions and work toward permanent solutions to problems, especially in cases where decisions will negatively affect the health and welfare of the natural world and future generations. As a result, cleanup levels must protect not only human health but also the health of the most sensitive species. Decisions must not only consider the cost of cleanup but also the rights of the natural world and our future generations. Native peoples, unlike most Americans and Canadians, are not a transient population and cannot abandon their homeland to find cleaner air, water, and land. Mohawk people, like resident species of plants and animals, will live adjacent to these Superfund sites forever and experience the effects of persistent contaminants for generations to come.

In Akwesasne, the goal is not only to permanently remediate these sites but to restore the land, water, air, and natural resources to their previous condition. The greatest legacy that the Mohawk people can leave their future generations is a healthy natural world. However, without adequate, permanent remediation, restoration efforts will be extremely difficult. Scientists such as Ward Stone, the wildlife pathologist for the state of New York, have confirmed fears that inadequate remediation will result in continuing damage to Akwesasne. Arguing that state and federal agencies respond much too slowly to environmental problems and do not have the will and support to enforce their own decisions, this prominent scientist has testified that:

ALCOA is still putting PCBs into the Grasse River. And they are going into the nearby Saint Lawrence into Indian waters. . . . Reynolds Metals is heavily hitting the River with PCBs, fluorides, aluminum, a wide variety of pollutants. . . . It's ongoing. The water is still being degraded. When I arrived in June 1985, the United States Environmental Protection Agency was involved at General Motors. . . . We still have a pollution problem more than five years later. The river is not cleaned up. EPA didn't identify Reynolds. It was the Pathology unit of DEC that identified that. . . . [T]he EPA is involved, but exceedingly slowly. And it's quite questionable whether or not their standards of cleanup will be sufficient to bring the river back—we think that it won't—to a place where the Mohawks will once again be able to utilize the fish and wildlife for food.[9]

At Akwesasne, decisions made by U.S. federal and state agencies, as well as those made by local industries, are preventing the Mohawk people from fulfilling their duties to the natural world and their people. What happens when two cultures clash on determining the future? Whose values and belief systems are respected? In this case, the decision-making authority of local industries and federal agencies such as EPA has directly conflicted with that of the Mohawk people. In balancing the interests of industry, the U.S. government fails to fulfill its responsibilities to protect the Mohawk people and the natural world. The initiation of preventive measures in protecting indigenous cultures requires a critical understanding of what is truly at stake.

Impacts: Extending Current Definitions

Like many indigenous peoples, the Mohawk people of Akwesasne have close ties to local ecosystems. They live, work, hunt, fish, farm and gather downstream, down gradient, and downwind from local industries. Even before the initial results were released on contaminant levels in area fish and wildlife, there was much concern regarding pollutants found in the local environment. Many residents, particularly women, were very concerned about possible health effects caused by toxicants. They even questioned the safety of breastfeeding their infants, especially because they unknowingly ate contaminated fish in the past. Research has shown that the developing fetus and infant are extremely vulnerable to chemical exposure and possible health effects. Many mothers found the thought of prenatal exposure to toxicants or exposure through breast milk very frightening. The anxiety they experienced from the fear of passing contaminants on to their infants was very real. This prompted Katsi Cook, a Mohawk midwife, to initiate discussions with scientists to perform chemical analysis of the breast milk of nursing mothers.

Over the past decade, Mohawk scientists, with the help of state and university researchers, have been conducting a full toxicological analysis of ecosystems at Akwesasne, including analyzing mammals, turtles, frogs, waterfowl, fish, invertebrates, vegetation, sediment, water, air, and human serum samples for contaminants such as PCBs. In July 1986, as a response to results indicating high levels of PCBs in area fish and wildlife, the St. Regis Mohawk Environmental Health Department issued an advisory to discourage Mohawk people from eating any local fish and wildlife in order to minimize potential adverse health effects. The advisory states that fish taken from the St. Lawrence River should be considered contaminated and not eaten by women of child-bearing age or by children under the age of 15. It also states that all people should eat no more than one meal (one-half pound) per week of fish from any body of water in or around the St. Regis Mohawk Reservation.

This advisory confirmed the fear of many residents, who had long been concerned that toxicants have contaminated surface water, soil, sediments, and air. Today, many parents do not permit their children to swim, fish, or play in rivers that may be contaminated. Anything grown at Akwesasne is suspect, even the air. Atmospheric fluoride levels have been recorded in the scientific literature as being so severe at Akwesasne that dairy cattle have been diagnosed as dying from clinical fluorosis.[10] There is much concern about children and elders who spend time outdoors, especially if they have respiratory conditions such as asthma.

Studies have shown that contaminant levels in the human population at Akwesasne have dropped dramatically as more people have refrained from eating local fish, wildlife, and vegetation.[11] Mohawk mothers who have historically eaten relatively large quantities of local fish have dramatically reduced their fish consumption over time, presumably as a result of the fish advisory.

This decrease in fish consumption, however, has prompted some scientists to make unwarranted conclusions. In speaking about Akwesasne, a prominent health researcher suggested that the community should be considered a success story because Mohawk people had eliminated their exposure to PCBs and other toxic substances. Mohawk people, however, have argued strongly that eliminating the consumption of local fish, wildlife, and plants is no solution to the problem of environmental contamination. Unfortunately, the voice of the community is not heard as often as that of prominent government scientists.

To suggest that Akwesasne is a success story is to suggest that it is acceptable for the victims of environmental contamination to continue to pay the price for pollution. Narrowly focused models of risk assessment make it almost impossible for scientists to understand the broader issues associated with persistent environmental toxicants. In classic models of toxicology, there is no risk if there is no exposure. Conventional risk assessments are severely limited in their application to native peoples because they fail to adequately value cultural, social, subsistence, economic, and spiritual factors. They also fail to consider issues of sovereignty, treaty rights, and self-determination.[12] The Mohawk people, on the other hand, have a much broader understanding of health and risk and have challenged the classical risk assessment models. They clearly have a different process of making decisions, one that is very precautionary in its approach.

Sometimes the greatest health effects are seen without any exposure and thus would not be included in risk assessments. At Akwesasne, human health has been affected by toxicants even without the ingestion of fish, wildlife, or water. For example, Mohawk people have customarily relied heavily on fish and wildlife as low-fat sources of protein, vitamins, and other important nutrients. Many health care providers at Akwesasne fear that the rapid changes in diet associated with the fish and wildlife advisories may be leading to diet-related illnesses such as heart disease, hypertension, stroke, and diabetes. The loss of fish in their diets represents the loss of an excellent source of protein. Thus, many Mohawks must turn to unhealthy, high-fat sources of protein, such as those found in fast-food places. Recent reports indicate that diabetes is on the rise because more people no longer eat traditional foods and no longer participate in cultural activities that once provided healthy forms of exercise.

In addition to the loss of protein, there is also concern about the use of herbal medicines. Air pollution from area industries has caused some traditional health care providers to question the use of locally grown plant medicines to treat sick people. Many elders rely heavily on traditional medicines and without the availability of local medicines, they fear a further deterioration of their health.

The effects of pollution, however, go well beyond physical health to include effects on economic, social, psychological, cultural, spiritual, and community health. The effects on the local subsistence economy, which includes important cultural and social institutions, have been particularly severe. The number of people who use local fisheries is difficult to estimate. However, almost every family in Akwesasne has at some time been involved with subsistence fishing. Before the ecosystem was contaminated with toxicants, some estimate that over 50 percent of the economy at Akwesasne was associated with fishing.[13] Today that number has dropped drastically. Unemployment in the community tends to be very high, averaging about 45 percent.[14] Sudden shifts in the economies of native communities can have devastating social and cultural effects.[15] Sudden changes in the environment have also been difficult for the people of Akwesasne.

When jobs are lost in the subsistence economy, Mohawks have to either leave their homeland to look for work or else promote forms of economic development which may not be compatible with cultural values or environmental mandates. As long as the community, its economy and the natural world are healthy, Mohawk people can maintain their independence and decide their own future. Environmental contamination severely limits the options available to the people of Akwesasne and few choices remain to support a stable, sustainable, and culturally appropriate economy. Industries have limited the ability of Mohawk people to determine their own destiny, which severely impacts their self-determination and self-governance. Discussions about the economic future of the community have become increasingly divisive because so few options are left. . . . The social and cultural destruction caused by environmental contamination impacts their very survival as a Nation.[16]

Environmental contamination threatens the well-being of entire nations of unique peoples in many ways and can be devastating to endangered cultures. It is often very difficult for non-native people to understand why environmental pollution results in cultural destruction and community breakdown. The only place the people of Akwesasne can be Mohawk is on Mohawk land. This in part explains why the people of Akwesasne have always viewed the protection of their land, air, and water as central to their social, political, economic, and religious structures. Subsistence activities are critical not only for economic stability but also for the long-term survival of the Mohawk language and culture. Traditional people are well aware that human existence is not possible without the natural world.

For native peoples, contamination of the natural world is an assault on their way of life. To explain all the ways in which toxic substances affect the Mohawk people requires an intimate understanding of Mohawk language, culture, history, and generations of traditional ecological knowledge. Loss of important cultural activities disrupts their family life and the extended family network. Fathers and sons are no longer able to fish together. Skills diminish along with the language associated with culturally important activities because fewer young people have the opportunity to learn them. Generations of traditional environmental knowledge are in danger of being lost.

Contamination of the natural world also has many psychosocial effects. The Mohawk people have felt enormous anger, sadness, and distress upon learning that dangerous levels of PCBs have been found in mother's milk as a result of eating fish.[17]

Not only do contaminated ecosystems result in a loss of peace and security. For Native people, pollution problems also result in lost relationships with the natural world, something that can only be likened to mourning. The people actually mourn the loss of the natural world. At the same time though, the ecosystem mourns the loss of Mohawk people, communities that have always worked to maintain a balance and restrict human activity to ensure the survival of all species. The loss of place, relationships and balance can be culturally devastating.[18]

Scientists, policy-makers, and the general public must reject the classical risk assessment models that not only constrain our thinking but that continue to institutionalize a form of environmental racism that destroys indigenous cultures and the land and resources upon which they depend for survival. The people of Akwesasne have strongly disagreed with the idea that decreasing consumption of fish and wildlife is the solution to the health risk associated with exposure to environmental toxicants. Although the body burdens of toxic substances may decrease, fish consumption advisories have many far-reaching effects.

Resisting Pollution: Mohawk Models Based on Respect, Equity, and Empowerment

Because of the many effects that pollutants have had on the Mohawk people, the story of Akwesasne, while not successful, is, however, a story of resilience, resistance, and hope. Since becoming aware of the contamination problems facing the community, the Mohawks of Akwesasne have voiced concerns not just about human health but also about the health and survival of the natural world—for the health of our future.

Because of this concern, the people of Akwesasne have mounted a strong response to environmental pollution and the destruction of their lands and waters. Motivated to provide a healthy, clean environment for future generations, they have vocally expressed their dissatisfaction with inadequate cleanup plans. The environment divisions of the Mohawk Council of Akwesasne and the St. Regis Mohawk Tribe (the two community governments of Akwesasne) have been working for years to address the issues of environmental contamination and promote adequate remediation of contaminated sites. However, it has been a struggle to participate effectively in the remediation process. To participate at all, they have had to find their own scientists and consultants, conduct independent research, and force the issue every step of the way.

The Traditional Mohawk Nation Council of Chiefs has also been actively involved in environmental issues at the international level and has sent delegations to the United Nations (UN) to remind the world that we as human beings have a responsibility to act as caretakers of the natural world. The Mohawk Nation has supported the development of a comprehensive plan to protect and restore natural resources and has worked with other Haudenosaunee nations to develop and present to the UN in 1995 a document entitled *Haudenosaunee Environmental Restoration: An Indigenous Strategy for Human Sustainability.*[19]

At the grassroots level, the Akwesasne Task Force on the Environment (ATFE), formed in 1987, addresses the environmental problems facing Akwesasne and proposes sustainable alternatives. Its members include the Mohawk community and the staff of environmental agencies and organizations within Akwesasne who share a common concern for the environment and the effects of toxic substances on human and ecosystem health. The ATFE works to conserve, preserve, protect, and restore the environment and natural and cultural resources within the territory of Akwesasne and has been involved in many projects that promote sustainable, culturally important activities.

Many ATFE projects concern environmental contamination. One such project is aquaculture (fish farming). Two ATFE aquaculture projects provide "clean" fish to local residents. Although no one at Akwesasne would ever consider aquaculture a permanent solution or a replacement for adequate remediation of contaminated sediments and on-site pollution, it offers a temporary solution that may allow the skills associated with fishing to continue. In time, it may also supply healthy protein to the community until full remediation and restoration occur.

Other projects sponsored by ATFE include restoration of indigenous fruit, nut, and sugar maple trees; seed giveaways; restoration of traditional medicines; and numerous workshops on natural resource management. To offset some of the negative cultural effects, such as a loss of traditional skills related to fishing, ATFE has also co-sponsored courses to teach adolescents these valuable skills with the aid of local fishermen and trappers. The project titled "Life Skills on the Land" helps to educate younger community members in these disappearing skills. The ATFE has also worked with community educators at Akwesasne to develop a culturally appropriate environmental education curriculum. Their goal is to ensure that young Mohawk students are taught in a culturally relevant way, so they learn the importance of environmental issues.

Finally, health studies at Akwesasne use an innovative research paradigm that is respectful, equitable, and empowering to the community.[20] The people of Akwesasne have strongly protested any research that considers native people as simply objects to be studied, and have worked to ensure that research conducted at Akwesasne supports sustainable science. Researchers are encouraged to involve Mohawk people not only in conducting research, but in designing useful and culturally appropriate projects. This new research paradigm allows information and science education to stay in communities even when projects end.

The First Environment Research Projects, for example, are part of a five-year study being conducted by Akwesasne field staff in conjunction with the State University of New York, Albany (SUNY) to examine the possible health effects of exposure to environmental contaminants. The first project is investigating the physical growth, maturation, and cognitive-behavioral development of adolescents at Akwesasne. A second project is examining how exposure to PCBs has affected the physical and mental health of adults. A third project is evaluating exposure to PCBs, measuring body levels, and assessing the overall health of the Mohawk community. A fourth project is focusing on grandparents to assess their PCB levels and exposure. A fifth project, the First Environment Communications Project, is designed to develop and implement communication strategies concerning the environmental studies. It helps provide the community with information on how to cope with contaminants by using culturally relevant strategies. For example, an alternative proteins workshop provided information and training on how to prepare meals using alternative sources of proteins.

Conclusion

The Mohawk community of Akwesasne resists change on many levels. Perhaps this resistance is reflective of the resilience of Mohawk culture, which has withstood centuries of attack. Although the Mohawk people have learned to adapt, they clearly cannot accept the environmental degradation that destroys the foundation of their culture. Toxic substances have polluted Mohawk lands and resources and helped to destroy their subsistence economy and traditional way of life. If this pollution is allowed to continue, if these toxicants are not adequately removed, and if natural ecosystems are not fully restored, then pollutants will greatly limit the options available to the community. The Mohawk people will potentially become just like mainstream Americans. Inadequate, temporary cleanups will serve as part of the machinery that acts to fulfill the longstanding attempt of the U.S. government to destroy native nations and assimilate native peoples.

By basing their decisions on mainstream values and scientific methods, government agencies deny the voice of native peoples and limit the decision-making authority of many sovereign native nations,[21] including the Mohawk Nation territory of Akwesasne. What is defined as an acceptable risk to most Americans is not acceptable to many traditional people, who do not participate in and value modern industrial economy in the same way.

Remediation decisions based on risk assessments and cost-benefit analyses rely on uncertain methodologies that are not value free. A risk assessment is based on assumptions about economic, environmental, and social priorities. Native nations certainly have different ways of making decisions about the natural world and these differences must be acknowledged, respected, and incorporated into any decisions that affect their future. If the U.S. government truly stands behind its laws and executive orders, it will recognize and respect the decision-making authority of Mohawk governments, including the Mohawk Nation Council of Chiefs, who have practiced democratic traditions that incorporate the protection of the natural world long before non-native people put foot on the North American continent.[22]

Economics inappropriately influences many remediation decisions. Permanent cleanups and restoration efforts that would ensure the survival of the Mohawk people are believed to "cost too much." Polluting industries are always quick to indicate how much they have already spent trying to clean up. With the constant fear of litigation and with a seemingly hostile Congress, federal agencies such as the EPA, an agency founded to protect the environment, hesitate to require the industries to permanently remediate sites. Apparently the EPA is weary of setting a precedent that would be applicable to all other Superfund sites. Who pays the price for these decisions? The effects caused by these contaminants may not always be reversible. What will the cost be for traditional skills and cultural practices lost due to the flagrant poisoning of our environment?

Mohawk people cannot remain passive as outside governments make decisions that will determine the future of Akwesasne, knowing that a healthy ecosystem is not only the key to a healthy community, but is essential for survival. For this reason Mohawks fight for cleanup and restoration of the natural world. Simply put, restoration of our mother, the earth, means restoration of Mohawk culture. As stated by the Akwesasne Task Force on the Environment:[23]

In order to understand the importance of adequate remediation, government agencies must learn to value the ecosystem from a Mohawk perspective. This is not going to be easy. Conflicting views on how the St. Lawrence River should be utilized are rooted in the socio-cultural, spiritual and historical relations that different cultures have with the land, water and the rest of creation. It is ironic that western science is just beginning to fully appreciate the scientific and cultural knowledge of Native peoples. At a time when western cultures are looking for models of sustainability, government agencies are simultaneously failing to protect Native communities and their lands. Native cultures continue to be oppressed and destroyed. We must ask if the American public agrees that Native cultures are expendable. We must also ask who is willing to stand up and protect the most vulnerable. What if the most vulnerable are unique cultures, irreplaceable lands, languages bordering on extinction, unborn children, and sensitive species of plants and wildlife? If Native cultures and traditional values are to be respected and given their rightful place, then the people of the world must be willing to give special consideration to protect those irreplaceable cultural resources.

Many native peoples have argued that one of the most important gifts that they have to offer the world is their understanding and appreciation for the natural world. If they are to be able to share this gift, then their cultures and the environments on which they depend must be protected. This must become a high priority, since with every passing day, more knowledge and cultural diversity are lost. Today, more than ever, there appears to be a real need for an environmental ethic in mainstream society, whose culture apparently lacks any particular love or affection for the natural world. In order to build this ethic, however, we must first acknowledge and affirm the rights of all living things to continue their existence. Mohawk traditions speak of the need to love and appreciate our relatives in the natural world—the earth, grasses, trees, birds, animals, medicine plants, waters, crops, thunders, sun, moon, and stars.

The people of Akwesasne are proud to be part of the continuing battle to determine whether life will be possible. Whose vision of the world will win—the voices of creation or the voices of destruction? We can only hope that the activism and science at Akwesasne will support other communities who continue to fight to ensure that responsible, long-term, equitable, sustainable environmental solutions are found that will protect all of creation, so that life may continue.

Notes

1. J. Sunday, *The Community of Akwesasne* (Mohawk Territory of Akwesasne. Department of Environment, Mohawk Council of Akwesasne, 1996); Mohawk Council of Akwesasne, *Akwesasne Mohawk Council Resolution for the Protection of Existing, Contemporary, Historic and Ancient Burial and Historic Sites and Their Remains and Contents* (1996); Mohawk Nation Council of Chiefs, "Resolution to Support a Conjoint Community Law for the Protection of Existing, Contemporary, Historic and Ancient Burial and Historic Sites and Their Remains and Contents, within the Akwesasne Community and within the Akwesasne Mohawk Land Claim Areas of the United Mohawks of Akwesasne" (September 23, 1996).

2. Akwesasne Notes, *Basic Call to Consciousness* (Summertown, Tenn.: Book Publishing, 1984).

3. C. Jacobs, "Presentation to the United Nations," *Akwesasne Notes* 13(4) (1995):116–117.

4. F. H. Lickers, "Presentation to the United Nations," *Akwesasne Notes* 13(4) (1995):16–17.

5. Lawrence Hauptman, "Drums Along the Waterways: The Mohawks and the Coming of the St. Lawrence Seaway," in *The Iroquois Struggle for Survival,* Lawrence Hauptman, ed. (Syracuse, N.Y.: Syracuse Univ. Press, 1986).

6. Akwesasne Task Force on the Environment, R.A.C. "Superfund Clean-up at Akwesasne: A Case Study in Environmental Injustice," *International Journal of Contemporary Sociology* 34(2), (1997):267–290.

7. St. Regis Mohawk Tribe Environment Division, press release (June 6, 1995).

8. R. Bullard, "Anatomy of Environmental Racism and the Environmental Justice Movement," in *Confronting Environmental Racism: Voices from the Grassroots,* R. Bullard, ed. (Boston: South End Press, 1993).

9. Ward Stone, "New York State Assembly Hearings: Crisis at Akwesasne, Day II, Transcript," in D. Grinde and B. Johansen, *Ecocide of Native America* (Sante Fe, N.M.: Clear Light Publishers, 1995, p. 195).

10. L. Krook and G. A. Maylin, "Industrial Fluoride Pollution: Chronic Fluoride Poisoning in Cornwall Island Cattle," *Cornell Veterinarian* 69 (Suppl. 8) (1979): 1–70.

11. E. Fitzgerald, S-A. Hwang, K. Brix, B. Bush, K. Cook, and P. Worswick, "Fish PCB Concentrations and Consumption Patterns Among Mohawk Women at Akwesasne," *Journal of Exposure Analysis and Environmental Epidemiology* 5(1), (1995):1–19.

12. B. L. Harper, "Incorporating Tribal Cultural Interests and Treaty-Reserved Rights in Risk Management," in *Fundamentals of Risk Analysis and Risk Management,* V. Molak, ed. (Boca Raton, Fla.: Lewis Publishers, 1997); S. A. Curtis, "Cultural Relativism and Risk-Assessment Strategies for Federal Projects," *Human Organization* 51(1), (1992):65–70; Shoshone-Bannock Tribes, "Risk Assessment in Indian Country: Guiding Principles and Environmental Ethics of Indigenous Peoples," position paper written as part of the National Tribal Risk Assessment Forum, "An Indigenous Approach to Decision Making," 1996.

13. Author unknown, Draft Report, Public Involvement and Case Study of the General Motors Superfund Site (Massena, N.Y., 1987).

14. Sunday, *Community of Akwesasne,* note 1.

15. P. Kettl, "Suicide and Homicide: The Other Costs of Development," *Northeast Indian Quarterly,* 84 (1991):58; A. M. Shkilnyk, *A Poison Stronger than Love: The Destruction of an Ojibwa Community* (New Haven, Conn.: Yale Univ. Press, 1985).

16. Akwesasne Task Force on the Environment, 1992.

17. J. Johnson, "Mohawk Environmental Health Project Integrates Research into the Community," *Environmental Science and Technology* 30(1) (1996):20A.

18. Akwesasne Task Force, note 6.

19. Lickers, "Presentation to the United Nations," 1995.

20. Akwesasne Task Force on the Environment, R.A.C. "Akwesasne Good Mind Research Protocol," *Akwesasne Notes New Series* 2(1), 1996:94–99.

21. Tom Goldtooth, "Indigenous Nations: Summary of Sovereignty and Its Implications for Environmental Protection," in *Environmental Justice: Issues, Policies and Solutions,* B. Bryant, ed. (Washington, D.C.: Island Press, 1995); R. Tomosho, "Dumping Grounds: Indian Tribes Contend with Some of Worst of America's Pollution," *Wall Street Journal,* November 29, 1990; pp. A1, A6.

22. Akwesasne Task Force on the Environment, Superfund Clean-up at Akwesasne," 1997, note 6.

23. Ibid.

7

When Harm Is Not Necessary: Risk Assessment as Diversion

Mary H. O'Brien

Simple Morality Undone

Most humans retain a quite simple, human morality about harm: It is not right to inflict suffering unnecessarily on another living being. If, for instance, you asked people, "Is it right to unnecessarily harm a child?" the almost universal answer would be, "No." If you asked, "Is it acceptable to unnecessarily harm a hummingbird, or an elderly person?" most would answer, "No."

However, most humans also are remarkably creative at justifying both small and huge breaks with the simple decency of avoiding unnecessary harm to others. In such situations, we may avoid gathering information about the harm or suffering we are causing, or the harm we are allowing to go on. We avert our eyes. (Ask many victims of incest, for instance, if their mother tried to avoid recognizing signs of what was happening.)

Or we may fail to make obvious connections that would make our personal choices and decisions suspect to ourselves. (Ask wealthy shoppers, for instance, whether they have thought of why the conventionally grown tomatoes they are buying are cheaper than organically grown tomatoes.) Most important, we often simply avoid considering alternatives that would not cause harm.

Yet we tend to admire people who face up to the distance between their sense of decency and their less-than-decent actions. Remember the captain of the slaving ship, who wrote "Amazing Grace" after he recognized that slavery and the role he was playing in it were fundamentally wrong, and not necessary?

Our personal tendency to justify unnecessary harm and to avoid the sometimes wrenching change that is necessary to cease unnecessary harm becomes a fiercely defended way of life at corporate and societal levels. The "bottom line" is the ultimate justification for dewatering and poisoning rivers; making the world's wildlife and humans stupid, sick, and sterile with endocrine disrupters; drowning vast regions of life with boondoggle dams; ripping away the ozone layer by injecting a soil fumigant underground; and on and on, step by step.

And that's the trick. Step by step.

In a memorable op-ed piece, Stanley Fish describes how, in the first Rodney King beating trial, the police officer defendants were acquitted.[1] How, he asks, could that have happened, when the jury had seen the video of the beating? A part of the answer lies, he concludes, in the defense's effective strategy of showing each frame of the video one by one, and asking questions that treated each frame as a separate case. "Is this blow an instance of excessive force? Is this blow intended to kill or maim?"

Similarly, developers and loggers have denied "harm" to the endangered piping plover from destruction of their last remaining breeding grounds, saying that no single plover would have been targeted, and no living plover was injured. Likewise, with affirmative action: Those opposed to incentives for contractors to hire minority subcontractors say that remedies for discrimination should be allowed only in cases where there has been "a harm inflicted on a specific person by a specific agent at a specific time."

Fish asks why arguments like these have so much force, and concludes that the trick lies in deflecting the eye away from the whole, whether history, habitats, or culture; the parts can then be described in any way one likes. "But why," he asks, "is the sleight of hand successful? Why don't more people see through it?" "Because," he answers, "it is performed with the vocabulary of America's civil religion": the vocabulary of color blindness, individual rights, equal opportunity.

Let us look at a similar example in the realm of toxics. There is a large incinerator (called the WTI incinerator) in Liverpool, Ohio, that burns hazardous wastes. Because the incinerator does not completely burn all of the toxic chemicals and does not burn toxic metals that are present in the wastes, it emits some highly toxic chemicals and metals into the air. The incinerator has been located 400 yards from an elementary school. This school is up on a bluff, approximately level with the top of the incinerator stack. The toxic chemicals that leave the stack are carried in the wind over to the school, where the children are hit by them every day. They breathe the toxic chemicals into their lungs, they absorb them on their skin, they pick up some on their fingers by playing in the schoolyard, and then ingest some whenever they stick their fingers in their mouth or eat their food.

The incinerator did not have to be built here. In fact, the incinerator wouldn't have had to be built at all if we provided incentives and requirements for even modest efforts at reducing use of toxics, and encouraged recycling and reuse of materials.

How, then, does one justify the unnecessary operation of an extremely large, hazardous waste incinerator 400 yards from an elementary school? Answer: risk assessment. Defend the indefensible by concluding that no significant harm will be done. Dress up damaging activities in complicated formulas, cancer models, toxicological testing, and monitoring. Don't ask, "Is there something we could be doing that would avoid all of this harm?" Risk assessments encourage us to "See no evil, hear no evil, speak no evil."

Industry consultants and government employees have calculated the amount of toxic chemicals and metals that WTI, the ever-shifting and re-forming corporation that runs the incinerator, can dump on the elementary school children.[2] They have decided, for instance, that the incinerator managers can annually send 9,400 pounds of lead, 2,560 pounds of mercury, and 157,400 pounds of fine particles up and out of the 150-foot incinerator stack.[3]

The unnecessary incinerator, the poisoning and suffering are justified by deflecting the eye away from the whole. Risk assessment effects that deflection by using the technical vocabulary of science: experiments, data, extrapolation, numbers, formulas. After all, can you prove that 157,400 pounds of fine particles released from the incinerator stack have harmed or will kill any given child in the elementary school?

In this chapter, I describe the basics of risk assessment; its failure to deliver results other than suffering and environmental degradation when it is applied to toxic substances; some of the reasons for its powerful hold on our society, our lives, and our deaths; and how to rescue life from it. This chapter is not an academic exercise because, as the primary process to justify pollution, risk assessment will not be dethroned on the basis of discussions among scholars. This will depend on the commonsense and conscious resistance of citizens, on the basis of simple morality.

Cover for Harm

What is toxic risk assessment? If it were primarily the consideration of potential deaths, illnesses, and degradations that given toxics could wreak, as the phrase "risk assessment" seems to imply, it would be useful. If this assessment of potential havoc were separate from the intent to defend toxics, it would be sensible, but it is not.

Environmental risk assessments are overwhelmingly used to construct a scientific-looking claim that an unnecessary, hazardous activity or substance is "safe," or poses "insignificant harm," or, when dead bodies clearly belie this, is "acceptable." Environmental risk assessment is thus used to defend the indefensible: production of dioxin; incineration of plastics and nerve gas weapons; genetic engineering for herbicide resistance in crops; fueling of cement kilns with hazardous waste; spreading of industrial, hazardous wastes on agricultural land; feeding cows with dead animals and injecting them with bovine growth hormone. Propose it, hire a risk assessor, and obtain the permit.

I concentrate here on the risk assessment of toxics, but, as Stanley Fish showed, the same risk assessment strategies can be used to excuse each blow in a police beating; each new house built in a piping plover's breeding grounds; or each hire of a white friend instead of a black stranger.

In the realm of toxics, risk assessments serve as the basis of social permits to distribute toxics into public places: for instance, to register, sell, and spray pesticides; pollute drinking water with toxics (Maximum Contaminant Levels); pollute rivers with toxics (Total Maximum Daily Loads); contaminate food with pesticides (Acceptable Daily Intake) and air with incinerator toxics; and emit solvents and other toxics into workplace air (Threshold Limit Values). Most regulation of toxics in our society is based on permitting the use and release of toxic substances in amounts that the producers and users claim are essential for them to sell, use, and release. Risk assessment provides a "regulation" cover for this use and release of toxics via its pronouncement that these amounts will be safe or not too harmful. In reality, "regulation" of the 70,000 chemicals in commerce is almost nonexistent, except in the case of a few high-profile chemicals whose releases, at least to the air or water, have been limited somewhat after much expense and fanfare; and the extremely rare case in which a chemical is actually phased out.

Defenders of toxics will claim that the basis of risk assessment is the reality that "the dose makes the poison"; i.e., that every substance, even water or table salt, will be harmful at some dose, and will be safe (or not extremely harmful) at a lower dose. Thus, according to this reasoning, there is no useful distinction between toxic and nontoxic substances. Each substance merely awaits its risk assessment prince to kiss it and make it beautiful. Therefore, risk assessment methodology is based on two assumptions:

1. "Some dose of the poison is safe" (or the variant for toxics like carcinogens, of which one molecule may pose risk, "Some doses of some poisons aren't too harmful").

2. "We can figure out which dose is safe or not too harmful."

The spectacle of scientists and risk assessors calculating a safe or insignificantly harmful dose of a toxic is set up on a stage by those who wish to defend the use of the toxic substance. The performers employ expensive toxicological experiments, intricate formulas, big computers, lots of paper, and explosive debates. (Often the performers do not realize they are being set up on a stage.)

Meanwhile, the defenders of toxics hope that the spectacle on the risk assessment stage that they are helping to fund and mandate through legislation will be so compelling and large that no one will wander off to the other, smaller stage, where alternatives to the use of the toxic substance are on display.

The stage for alternatives is generally quite small, and the displays are often homegrown and always underfunded, but it is the alternatives that hold the promise of environmental restoration, social change, and democracy.

The Basics of a Risk Assessment

The production of a risk assessment for a given toxic substance can be informal or complex; understandable or inaccessible to all but a small cadre of statisticians and biometricians; brief or ponderous; back-of-the-envelope, or costly to seven figures. The basic process of producing many types of toxic risk assessments, however, has three steps:

1. *Estimate toxicity and lack of toxicity.* Gather some data on some of the consequences of exposing a living organism to some toxic substance.

Generally, such data are gathered in a laboratory environment in which all factors are kept constant in the control and experimental groups of animals (or other living organisms). Only the exposure to the toxic chemical varies. The experimental animals are exposed to a range of doses of the toxic chemical, while the control group is not exposed.

Sometimes data are gathered out in the world, for instance from human epidemiological surveys (health damage) or surveys of wildlife or vegetation damage. An attempt is made to find an experimental or field dose below which the toxic effects are not seen (i.e., a No Adverse Effect Level). This first step will provide some information about some type of toxicity of the substance, but as indicated below, there are many ways a substance might be toxic).

2. *Estimate expected real-world exposure.* Estimate, with greater or lesser amounts of data, the exposure to this chemical that certain organisms (e.g., workers, the general public, birds, plants) will be expected to encounter as a result of permitted production or use or release of this toxic chemical.

Such estimates are variously simple or complex, depending on whether the assessor is going to estimate, for instance, how much of a pesticide will land on a person's skin, how much of the person's skin will be exposed, how much of the pesticide will be absorbed into the body, how much will land on food plants, how long pesticide residues will remain on the plants, how many vegetables or fruits the person will eat in how much time, how much will land in water, how quickly the pesticide will break down in water, how much water the person will drink before the pesticide has broken down, and how much the person weighs.

3. Compare threshold or potency of toxicity with expected exposure. For those substances that appear to have a threshold dose below which the particular adverse effect appears not to occur, compare the estimated level at which toxic effects can first be expected, with the expected exposure. This is done to estimate whether the expected exposure will be safe or not too harmful.

For those substances that appear to be capable of causing damage (e.g., genetic damage, cancer) with even one molecule, multiply the expected exposure by the potency of the substance in order to estimate the chances that the damage will occur.

Notwithstanding the necessity to make estimates, this seems on the surface to be a straightforward scientific task. It is not, for many reasons that have been exhaustively examined in hundreds of books and thousands of journal articles. Two major factors account for the task not being straightforward. The first is scientific: We cannot know all the ways a substance might be toxic and therefore cannot estimate what exposure will be safe or not too harmful. The second is political: Those whose money or job lies in ensuring the production, sale, use, and disposal of the toxic substance will try to make sure that the risk assessment does not indicate harm (i.e., that it will "See no evil, hear no evil, speak no evil").

Too Many Ways of Being Toxic

In order to determine that some dose of a toxic substance will not cause harm, a risk assessor has to know that all of the ways the substance could be toxic have been examined. However, this is not possible.

First, consider just some of the ways a substance can be toxic:

• It can cause:

Cancer (e.g., testicular, lung, ovarian, breast, brain, stomach, bladder, skin, bone, cervical, and a host of other cancers)

Birth defects

Immune system suppression (e.g., by disabling T-helper cells or killer cells, or a host of other ways)

Skin irritation

Genetic damage (e.g., point mutations, sister chromatid exchanges, or a host of other kinds)

Respiratory disease

Kidney degradation

Liver damage

Nervous system disruption (e.g., slowed nerve conduction, peripheral neuropathy, cholinesterase inhibition, or a host of other effects)

Sterility

Spontaneous abortions

Reduced birth weight

Reduced head-to-body ratio at birth

Reduced intellectual functioning (e.g., loss of short-term or long-term memory)

Emotional instability (e.g., anxiety, depression)

Hormone disruption (e.g., estrogen mimicry, reduction of testosterone production, inhibited production of insulin, or a host of other disruptions of the 100 or so hormones present in the human body alone)

• Moreover, a given substance can interact with another substance in a synergistic manner to produce heightened toxicity (e.g., one plus one equals thirteen rather than two). Exposure to one substance can add to the toxicity of exposure to other substances (i.e., one plus one equals two).[4]

• A chemically sensitive person may be hypersensitive to a substance that appears to not affect most people, even at much higher doses.[5]

• A toxic substance may break down in the environment or be transformed into other substances that are more toxic than the parent substance or toxic in a different way.

• A toxic chemical may be nontoxic at high doses, but cause damage at doses far below those tested in standard laboratory experiments.[6]

Most of the 70,000 chemicals in commerce have not been even minimally tested. For instance, the Environmental Defense Fund (EDF), a large nongovernmental organization (NGO), recently examined the testing information that is publicly available for a random sample of 100 industrial chemicals that are produced in high volume, and have also been identified as subjects of regulatory attention under major environmental laws. You might expect that these, of all chemicals, would have been examined for their toxicity. EDF looked to see if these high-profile, high-volume chemicals have had even a minimal amount of toxicity testing [i.e., tests for acute (immediate) effects, repeated dose (chronic or more long-term) effects, genetic damage (tested in a test tube and in laboratory animals), reproductive damage and developmental damage]. They found that about 75 percent of these chemicals lack even this minimal information, never mind information about immune system effects, nerve damage, hormone disruption, synergism, or breakdown products.[7]

So how does a scientist or risk assessor claim, with any seriousness, to have established exposure to a particular toxic chemical as safe or minimally harmful? It is only possible to state, with scientific integrity, some of the harms that might be caused by a chemical. It is not possible to indicate that exposure to the chemical, in conjunction with other chemicals, will *not* cause harm. Yet the supposed demonstration of no harm or minimal harm is the raison d'être of risk assessment. It is how the chemical industry, agribusiness, corporations, government permitters, and international trade associations justify unnecessary harm and stave off change.

Too Many Openings for Politics

The absence of data rarely stops a risk assessor, however. (In fact, data can get in the way.) If only one toxicity test has been performed, no matter how poorly, the risk assessment will use numbers from that study. If a particular type of study has not been done, for instance of how the substance is absorbed through the skin, the risk assessor will rely on a number from a study of some closely or distantly analogous substance. This, of course, requires selective judgment by the risk assessor and her or his employer.

Sometimes the risk assessment does not provide good news: The substance appears to pose the danger of ill effects. Assumptions may then be changed. A decade ago I decided to cease spending time encouraging accurate risk assessments after I saw how readily the assessments were manipulated to produce desired answers. For instance, an herbicide risk assessment, using data from an extremely poor chronic effects study on animals, showed a risk of health impacts from the levels of the herbicide dacthal that had been found in eastern Oregon groundwater. The risk assessment was subsequently adjusted by the U.S. Environmental Protection Agency (EPA) to eliminate a factor in the assessment model that took into account the fact that children would be drinking the water, as opposed to only adults.[8] Now the health threat disappeared.

Many people know of William Ruckelshaus' famous description of risk assessment: "We should remember that risk assessment data can be like the captured spy: If you torture it long enough, it will tell you anything you want to know."[9] At the time he said this, Ruckelshaus was administrator of the EPA, and he was promoting the establishment and use of risk-benefit analysis within that agency.

A risk assessment is not reproducible because it shifts, depending on which numbers, assumptions, estimates, and formulas are used. When Intel's Pentium computer chip was found to be making some arithmetic mistakes, Intel's risk assessors determined that an average computer user would get a wrong answer once every 27,000 years of normal computer use. Risk assessors with IBM, which had a competing microprocessing chip to sell, estimated that an average user would get a wrong answer every once 24 days using Intel's Pentium chip.[10]

If risk assessors (with different interests) cannot agree on how often a computer chip will make an arithmetic error, they certainly will not (and do not) agree on how a toxic chemical will behave as it works its way through soils, air, water, plankton, turtles, and humans.

If wildly different estimates of harm can be (and are) produced through risk assessment formulas, then it matters who is at the table playing the game. Producers and users of toxics frequently have many millions of dollars and market shares riding on the sale and use of a particular toxic chemical. Public agencies, using public money, defend their permitting of unnecessary toxics and their own uses of toxics.

Citizens' organizations cannot win at this game, and should not spend their time trying, because the other players have too much incentive and the money to produce risk assessments that assure the public that their exposure to the chemicals will be safe. If 300 citizens' organizations working on toxics issues were assembled in one place to describe their individual horror stories of involvement in or subjugation to the consequences of risk assessments, they could compile hundreds of tales. Their stories would variously involve local county health departments, chemical companies, state departments of agriculture or environmental quality, the U.S. Environmental Protection Agency, U.S. Forest Service, U.S. Army, and many others. Each story would be a testimonial to the near-constant manipulations of risk assessments for the political purpose of demonstrating minimal harm or none from unnecessary use of toxic chemicals.

There are a number of reasons private industry aggressively promotes risk assessment:

• It is the basis of their permits to produce, sell, use, and dispose of toxics.

• It diverts attention from whether they need to produce, sell, use, or dispose of these toxics.

• It gives industry the aura of being scientific about the "safety" of their activities.

• The complexity of most risk assessments allows for interminable haggling, while toxics business-as-usual goes on. (How many years, and millions of public dollars, for instance, have been sunk into the past decade's EPA risk assessment of dioxin, the most toxic chemical humans produce? Meanwhile, the chlorine industry, the source of this dioxin, continues carting its profits to the bank.)

• Complicated risk assessments let industry's hirelings be the experts.

• It is far more convenient to reduce "risk" and damage on paper by writing optimistic risk assessments than to reduce real-world damage by ceasing hazardous activities and adopting more environmentally appropriate practices.

• The use of risk assessment to "harmonize" standards internationally (e.g., through the *Codex Alimentarius* for pesticide residues) is used to force countries to allow the importation of unsafe technologies or products under the banner of free trade [e.g., through the General Agreement on Trade and Tariffs (GATT) and the North American Free Trade Agreement (NAFTA)].

There are a number of reasons government agencies prepare and use risk assessments:

• Industry has written and promoted the passage of laws requiring many government agencies to write risk assessments.

• Risk assessment processes allow government permitters to hide behind "rationality" and "objectivity" as they issue permits and allow hazardous activities that harm people and the environment. After all, it is not easy to justify permitting a hazardous waste facility 400 yards from an elementary school.

• Risk assessments can be manipulated endlessly to accommodate desired policy.

• Since risk assessments are not reproducible as a scientific experiment or study, they therefore cannot be disproved. The agency will not likely be overturned in its decision-making.

• Risk assessment models and numbers are intimidating to citizens, thereby giving agencies the "upper hand."

• If agency employees can focus the public's attention on the details of a risk assessment, they will divert public debate from consideration of whether the assessed activity should even occur.

    So we have industry and government preparing and using risk assessments as the key justification for permitting continued production, use, and disposal of toxic substances. Could we be doing something else?

Do We Need Better Risk Assessments?

Harvard Center for Risk Analysis Director and risk assessment promoter John Graham disapprovingly notes that "[M]ost environmental advocacy organizations have resisted the use of risk analysis for moral, technical, and/or tactical reasons."[11] Those citizens' groups who do accept or even promote risk assessment generally fall into two camps. In one camp are newly formed grassroots organizations who have not yet had the inevitable experience of having the risk assessments they are working to influence be used against them. They may understand that the risk assessment they are addressing is being used to justify their exposure to toxics, but they may still believe that if they provide documentation that the chemi cals are more toxic than admitted in the assessment, the assessment will be changed and so will the decision to expose them to toxic chemicals.

In a second camp are large, national environmental organizations who can afford to hire staff scientists to urge improvements in bad risk assessments, and who use risk assessments to urge more regulation. Believing that risk assessment is the major playing field, they do not want to miss being a "player" in the game, along with corporations and government agencies.

John Graham approvingly notes, for instance, an "exception within the advocacy community": The Environmental Defense Fund, "which has an aggressive, forward-looking approach to the use of risk analysis in favor of environmental protection."[12] As seen earlier, the EDF is the same organization that has assembled information showing that less than 25 percent of one hundred high-profile, top-selling industrial chemicals have had even minimal toxicity testing. Despite the presence of even less information for most of the other 70,000 chemicals in commerce, EDF refers, in its "Scorecard" Web site[13] to "safety assessments" of chemicals. "For example," the Web site states, "the safety of releases of benzene to air can be assessed for cancer risks because EPA has an inhalation cancer potency value for the chemical."[14] So, can we determine the "safety" of a potent carcinogen, benzene, whose toxicity may have no threshold, by referring to the U.S. Environmental Protection Agency's cancer risk model for inhalation of benzene?

When I questioned EDF Scorecard author and toxicologist Bill Pease about "safety" assessments, he said, "I think you can conduct as good a safety assessment as is generally feasible if you force the generation of local exposure data and the development of toxicity values" (personal communication from B. Pease, September 20, 1998). In other words, he recommends pressing industry to test the 70,000 chemicals in commerce (for all the ways of being toxic?). He recommends that someone (who?) should estimate a community's exposures to each chemical, presumably after expanding reporting requirements for toxics. Currently reporting is required for only a fraction of all toxics used and released.

In reference to EDF's choice of the phrase, "safety assessment," Pease continued, "We needed to find a phrase that was publicly interpretable and that addressed the most typical way questions about chemical releases are expressed: 'Is it safe?' and the language the chemical industry is using to respond to these questions: 'It's safe'" (personal communication from B. Pease, September 20, 1998). Pease has here mentioned the core issue of this chapter. Is the "most typical" question, i.e., "Is this toxic chemical safe?" the one that we should be asking? I do not believe so. I believe it is the question industry has taught us to ask and the one they want to answer, using risk assessment.

The question, "Is this toxic chemical safe?" is akin to asking whether a given blow to Rodney King by a police officer was intended to kill or maim. Police beatings harm, and so do toxic chemicals. The appropriate question is, "Is this beating, or use of this toxic chemical necessary?" Our task is to install this question into the processes of permitting the production, use, and release of toxic chemicals.

A Question of Necessity

On the other hand, when we ask, "Is use of this toxic chemical *necessary?*" we move to a different playing field than the risk assessment field. We move to the alternatives assessment field because the question can only be answered by reviewing alternative materials and operating systems and comparing them with the toxic chemical.

Since no alternative can be called completely "safe" (for the same reasons no exposure to a toxic chemical can be declared "safe"), we can only compare the various alternatives for the types of benefits and problems each offers. The assessment of potential problems of each alternative might involve "risk" assessment (i.e., a discussion of potential health or wildlife hazards of toxic chemicals or other elements involved), but other types of problems (e.g., economic costs, lack of information about the alternatives' consequences, difficulty of monitoring) could also be considered problems along with toxicity.

Using this approach, the various benefits of all alternatives can also be considered, whether short term or long term. For instance one or more alternatives may be less expensive than use of the toxic chemical, or it may be more aesthetically pleasing. In such a situation, a toxic chemical may not fare well at all, even without information "proving" toxicity, or even if the substance is apparently only mildly toxic.

The defenders of risk assessment often claim that the consideration of alternatives is a complement to, not replacement of, risk assessment. However, the characterization of a toxic chemical examined in an assessment of alternatives does not require any estimate of a "safe" or "insignificantly harmful" level of exposure to the chemical.

The following statement on the noncomplementarity of risk and alternatives assessments was sent by e-mail in September 1998 to a group of Russian and North American scientists, NGOs, and government representatives that had been meeting in Siberia a month earlier to examine concerns about dioxin and other persistent organic pollutants (POPs) in Russia. The author of the e-mail message, Penny Newman, is a community member and grassroots activist in a Southern California town that is host to a 40-year-old Class I hazardous waste site containing more than 200 toxic chemicals. Newman was responding to a statement by a Canadian scientist in the group to the effect that risk assessment and alternatives assessment complement each other. (Copy of e-mail available from author.)

Contrary to what everyone tries to portray, risk assessment and alternatives assessment do not complement each other. They ask different questions and arrive at different and conflicting answers. The whole purpose of risk assessment is to come up with a magic number that purports to be a "safe level." From that point on, the focus is on reaching and maintaining that level, not in stopping exposure. To suggest that the problems aren't with risk assessments but in their "misuse and perversion" is like saying "Guns don't kill, people do." There is a fundamental flaw in risk assessments that makes them dangerous in any setting.

Let me give an example. In 1983, [PCE] was discovered in the drinking water fountains at the elementary school. The state conducted a risk assessment that concluded that children drinking this water were not at an unacceptable risk if the levels stayed below 5 ppm. They did not tell us that the assessment was based on estimated cancer outcomes, not non-cancer health impacts; that it did not consider the effects of PCE mixed with chlorination or other contaminants in the water; that it was based on a healthy adult male, not on children that had been exposed for years to more than 200 different compounds; nor did they outline all the other factors not considered.

During a break in the meeting where this assessment was being discussed, I asked the state official if he would let his little 3-year-old daughter drink that water. His response was, "Hell no!"

Our community rejected this approach. We couldn't understand why we should have to have our children drinking contaminated water at any level, especially since this was not naturally occurring, but the result of corporations dumping their wastes in our community. Wasn't it, therefore, their responsibility to correct the situation?

Instead of relying on a risk assessment, we did an alternatives assessment. We gathered all information known about the chemical and its health impacts, recognizing the limitations of that data (one can still analyze health impacts within this process without coming up with that magic number); reviewed available alternatives, and decided upon a filtration system for the school and a new drinking water system for the community. In other words, we didn't accept a "safe" level of pollution but pushed to stop the exposure completely. If we had relied upon the risk assessment approach, our kids would still be drinking contaminated water since the levels remained just below the "safe level."

Let's look at another example, this time in a book (*The Greening of Industry: A Risk Management Approach*) co-edited by John Graham, who was mentioned earlier in connection with the Harvard Center for Risk Analysis.[15] One of the chapters ("Cleaning Up Dry Cleaners" by Kimberly Thompson) provides a classic example of the extent to which our society will spend decades arguing about the precise risks of a toxic chemical, while (1) continuing to permit exposure of humans and other living organisms to the toxic chemical in workplaces, neighborhood residences, drinking water aquifers, and contaminated soil, and (2) avoiding the issue of whether use of the toxic chemical is even necessary.

The chapter focuses on a neurotoxic, mutagenic, and carcinogenic organic solvent called perchloroethylene ("perc"), which is used by the majority of the nation's 30,000 dry cleaners to "clean" clothes.

Arguments have raged since the 1970s about perc's ability to cause cancer in animals and humans; its relative contribution to formation of smog; its qualifications as a hazardous air pollutant and hazardous waste; its pollution of aquifers (which is not mentioned in the chapter); appropriate legal limits for worker exposure; and risks to residents in the neighborhood of a dry cleaner. The chapter traces intense disagreements between the EPA and its Scientific Advisory Board, which is heavily lobbied by dry cleaners' industry associations; litigation against worker protection from perc; the proliferation and challenges of animal experiments and human epidemiological studies; and regulations appearing after the industry realizes it will save money by reducing perc waste. Throughout these debates, the industry is constantly pressing for particular risk assessments (those showing less harm) and challenging other risk assessments (those showing more harm).

Toward the end of the chapter, however, the 1990s are reached, and mention is made of EPA's voluntary cooperative environmental design program called "Design for the Environment," which includes the development of "cleaner technology substitutes assessments" (CTSA), that is, alternatives assessment. A CTSA is "intended to help an industry evaluate pollution prevention oppportunities, and energy conservation associated with different solvents and technologies."[16]

Enter Greenpeace, promoting the use of water rather than perc as the solvent for cleaning "dry clean only" garments. When the results of this technology appeared promising, the CTSA process moved toward comparing various dry-cleaning alternatives. Tellingly, at this point the perc-based dry-cleaning industry suddenly resisted risk assessment. Industry leaders argued that the goal of the CTSA is to compare technologies and that numerical risk assessments might be harmful to the industry.

In a letter to the then-director of the Design for the Environment program, the Halogenated Solvents Industry Alliance argued, "This public display of supposed risks, especially if based on inappropriately simplistic assessment procedures, could harm the DfE program, embarass EPA, and unfairly harm the dry cleaning industry which is engaged in significant efforts to reduce [perc] releases to the environment"[17]

This is critically important to note, because by any possible comparative assessment of dry-cleaning technologies, water will look a lot better than perc. In the course of alternatives assessment, when the pros and cons of a reasonable range of alternatives are displayed, risk assessment suddenly loses its appeal for the chlorinated solvent industry. With the eyes of the audience on both stages, the chlorinated solvent stage and the water stage, arguments about whether the risks of perc are high enough to warrant regulation transform into arguments about whether perc is necessary at all. The issue thus changes from "What amount of risks are acceptable?" to "What options do we have for avoiding risks?" and those who are defending the risky option suddenly lose their stomach for risk assessment.

This is the key to understanding risk assessment: It is diversion when it is carried out apart from consideration of both the pros and the cons of a full range of alternatives. It is diversion from the essential question we must ask, both as moral persons and responsible societies: "What options do we have for treating each other, and the world, with care?"

Simple Morality Restored

In *Silent Spring*, Rachel Carson asks a startling question: "Have we fallen into a mesmerized state that makes us accept as inevitable that which is inferior or detrimental, as though having lost the will or the vision to demand that which is good?"[18] It is radical to return to the simple moral sense that it is not right to inflict suffering unnecessarily on other living beings. We wake up, as from a sleep, and necessarily have to look at all of our society's environmentally degrading activities (including economic growth and population growth) and materials, and ask which are necessary and which unnecessary.

There are sectors in our society that are doing just this. Organic farmers, to name one. While the organic farming movement does consider which toxic materials they will occasionally use to prevent excessive damage by pests, their main focus is on alternatives to pesticides and chemical fertilizers: rotation of crops, tilling, mulching, biointensive practices, cover crops, acceptance of pest losses, breeding plants for pest resistance, restoring the biological health of soils.

Organically grown food is a booming market, but it stands in stark contrast to another approach that is favored by the U.S. Department of Agriculture (USDA): genetic engineering. In December 1998, the USDA proposed standards for growing and marketing organically grown food. Contrary to current organic agriculture practices in the United States, and contrary to the recommendations of the National Organic Standards Board, the proposed standards allowed the use of genetically modified organisms (GMOs), and prohibited labeling any organically grown food with more information, such as "Not genetically engineered."

Why? Because current state organic farming standards stand as a viable alternative to the highly dangerous practice of unleashing genetically modified organisms on the world to increase crop yield, kill pests, or provide plant resistance to herbicides.

"Few if any existing [organic] standards permit GMOs and their inclusion could affect the export of U.S. grown organic product," reads an internal USDA memo written eight months prior to the release of the proposed organic agriculture standards.[19] The memo continues, "However, the Animal and Plant Health Inspection Service and the Foreign Agricultural Service are concerned that our trading partners will point to a USDA organic standard that excludes GMOs as evidence of the Department's concern about the safety of bioengineered commodities."

In other words, the official U.S. risk assessment of genetically engineered agricultural commodities has declared that genetically modified organisms are "safe." Under free trade agreements, countries cannot refuse imports of commodities that are unsafe, unless the importing countries can successfully show them to be unsafe. The U.S. risk assessment has declared them safe, but an alternative commodities market (i.e., the organic foods market) that rejects genetic engineering because it is not safe challenges this contention.

The consideration of and switch to alternatives can clearly mean major changes in how we do business in the world. Currently, our society is involved in a crucial decision-making process; negotiations with 150 other countries for a treaty regarding certain toxics called persistent organic pollutants (POPs). The negotiations are aimed at planning for the reduction and/or elimination of twelve persistent toxic chemicals worldwide, including dioxins.

The major issue under debate is whether the POPs treaty will commit to "eliminating" these persistent chemicals or merely "reducing" them. The chemical for which an "elimination" decision will be the most challenging is dioxin, because elimination of dioxin production means phasing out most industrial uses of chlorine. (Dioxin is an unwanted by-product of the production and heating of thousands of chlorinated hydrocarbon compounds.) Many of the other POPs under consideration have already been banned or extensively restricted in many countries.

If the treaty calls for the "elimination" of all of the POPs under consideration, humanity will have taken a decisive moral stance against the unnecessary poisoning of embryos and the rest of the world. If, on the other hand, the treaty calls for merely "reducing" some of these chemicals, especially dioxin, to "insignificantly harmful" levels, then we will continue to poison wildlife and humanity throughout the world.

In this context, it is useful to recall the 1992 statement by the International Joint Commission for Great Lakes Water Quality, which calls for elimination of such persistent toxics: "We conclude that persistent toxic substances are too dangerous to the biosphere and to humans to permit their release in *any* [emphasis in original] quantity."[20]

If the POPs treaty proposes elimination of dioxins and other persistent toxics, we will need to examine a large range of alternatives to the use of chlorinated chemicals. Conversely, if the *first* step in the POPs process were one of closely examining all reasonable alternatives to the use of these chemicals, the likely outcome would be a decision to eliminate rather than reduce the chemicals. This is because, as the Montreal Protocol process for eliminating the production and use of ozone-depleting chemicals such as chlorofluorocarbons (CFCs) has shown us, alternatives do exist, and life goes on, better than before.

Hope

Rachel Carson dedicated her book on pesticides, *Silent Spring*, to Albert Schweitzer, and quoted him on the dedication page: "Man has lost the capacity to foresee and to forestall. He will end by destroying the earth."

That is a grim quote. Risk assessment, the current process by which we claim that exposure to toxic chemicals is safe, is for the most part not used to foresee harm, but to forestall change. If, however, we turn our backs on the assessment of how much toxic exposure is safe, and turn instead to the question of our options for least use of toxic chemicals, we can regain the capacity to foresee.

"The choice, after all," Carson wrote later in *Silent Spring*, "is ours to make. If, having endured much, we have at last asserted our 'right to know,' and if, knowing, we have concluded that we are being asked to take senseless and frightening risks, then we should no longer accept the counsel of those who tell us that we must fill our world with poisonous chemicals. We should look about and see what other course is open to us."[21]

By mandating full assessment of alternatives in our personal, community, national, and international settings, processes, regulations, and legislation, we will be much more likely to refuse to damage the world unnecessarily.[22] We will encourage ourselves to act on the basis of simple morality.

Notes

1. Stanley Fish, "How the Right Hijacked the Magic Words," Op-Ed, *New York Times,* August 13, 1995, p. E15.

2. Hazardous Waste Facility Approval Board of the State of Ohio. Written opinion and final order approving application for hazardous waste facility installation and operation permit. Case No. 82-NF-0589 (February 16, 1984). Also, U.S. Environmental Protection Agency. Letter from Robert Sussman, deputy administrator, to Terri Swearingen, Tri-State Environmental Council (July 16, 1993).

3. Peter Montague, "The Breakdown of Morality," *Rachel's Hazardous Waste News* No. 287 (May 27, 1992).

4. Dioxins and furans chlorinated at the 2,3,7, and 8 and additional positions on their carbon rings, for instance, add together like 1 plus 0.01 plus 0.5. If a furan that is half as toxic (0.5) as 2,3,7,8-TCDD (the most toxic dioxin), is present at 100 times the amount of another that counts as one, that is 1 plus 50 equals 51. See Mary O'Brien, "A Crucial Matter of Cumulative Impacts: Toxicity Equivalency Factors," *Journal of Pesticide Reform* 10(2) (1990):23–27.

5. Nicholas Ashford and Claudia Miller, *Chemical Exposures: Low Levels And High Stakes* (New York: Van Nostrand Reinhold, 1990).

6. Frederick vom Saal, Paul Cooke, David Buchanan, Paola Palanza, Kristina Thayer, Susan Nagel, Stefano Parmigiani, and Wade Welshons, "A Physiologically Based Approach to the Study of Bisphenol A and Other Estrogenic Chemicals on the Size of Reproductive Organs, Daily Sperm Production and Behavior," *Journal of Toxicology and Industrial Health* 14 (1998):239–60.

7. David Roe, William Pease, Karen Florini, and Ellen Silbergeld, *Toxic Ignorance: The Continuing Absence of Basic Health Testing for Top-Selling Chemicals in the United States* (New York: Environmental Defense Fund, 1997).

8. See first, U.S. Environmental Protection Agency, Health Effects Branch, Office of Drinking Water, *Health Effects Guidance for Dacthal* (prepared for N.Y. State Department of Health) (Washington, D.C.: EPA, 1982). Next, see U.S. Environmental Protection Agency, Office of Drinking Water, *Dacthal Health Advisory* (Washington, D.C.: EPA, 1987). The dacthal "story" is described in Mary O'Brien, "Testimony before the Interim Committee on Environmental and Hazardous Materials," Oregon State Senate, March 9, 1988.

9. William Ruckelshaus, "Risk in a Free Society," *Risk Analysis* 4 (1984):157–162.

10. Peter Montague, "The Many Uses of Risk Assessment," *Rachel's Environment & Health Weekly* No. 420 (December 15, 1994).

11. John Graham, "Risk-Based Environmental Advocacy," *Risk in Perspective* 6(7) (1998):1–4.

12. Ibid.

13. Environmental Defense Fund, "Safety Assessment of Toxic Chemical Releases," http://www.scorecard.org/env-releases/def/assess.html (logged on October 16, 1998).

14. http://www.scorecard.org

15. John Graham and Jennifer Kassalow Hartwell (eds.) (Cambridge, Mass.: Harvard Univ. Press, 1997).

16. Ibid., p. 120.

17. Letter from Peter Voytek, Halogenated Solvents Industry Alliance, to Jean Elizabeth Parker, EPA, Design for the Environment Program staff director, June 27, 1994, cited in Graham and Hartwell.

18. Rachel Carson, *Silent Spring* (Boston: Mass.: Houghton Mifflin, 1962).

19. Richard Reynolds, "Mother Jones Releases USDA Memo Detailing Plans to Gut NOSB Recommendations on Organic Standards," press release. *Mother Jones Magazine* (March 12, 1998).

20. International Joint Commission, Sixth Biennial Report on Great Lakes Water Quality (Windsor, Ontario, Canada, 1992).

21. Carson, *Silent Spring*, p. 244.

22. The premier example of such regulations and legislation: Regulations for Implementing the Procedural Provisions of the National Environmental Policy Act, which requires the consideration of all reasonable alternatives for federal agency proposals. See 40 *CFR* Parts 1500–1508 (1992).

8

The Ecological Tyranny of the Bottom Line: The Environmental and Social Consequences of Economic Reductionism

John Bellamy Foster

In recent decades environmentalists have directed a persistent ecological critique at economics, contending that economics has failed to value the natural world. Lately economists have begun to respond to this critique, and a rapidly growing subdiscipline of environmental economics has emerged that is dedicated to placing economic values on nature and integrating the environment more fully into the market system. However, the question arises—is the cure more dangerous than the disease? Does the attempt to internalize the natural environment within the capitalist market system—without a radical transformation of the latter—lead to a new empire of the economy over ecology, a sort of neocolonialism where the old colonialism is no longer seen as sufficient? And what are the ultimate consequences of this?

Although there are distinguished exceptions, most work in the relatively new field of environmental economics is conducted within the orthodox or neoclassical economic framework.[1] As the British left-green economist Michael Jacobs has written, "At heart, the neoclassical approach to environmental economics has one aim: to turn the environment into a commodity which can be analyzed just like other commodities. . . . If only the environment were given its proper value in economic decision-making, the economist reasons, it would be much more highly protected." "As far as economists are concerned," George Eads and Michael Fix likewise observed in a study published by the Urban Institute, "the problems of environmental pollution, excessive levels of workplace hazards, or unsafe consumer products exist largely because 'commodities' like environmental pollution, workplace safety, and product safety do not trade in markets."[2]

For orthodox economists, ecological degradation is evidence of market failure.[3] The market is unable to guide firms in the efficient use of environmental assets if they are not already fully incorporated within the market system by means of a rational price structure. The first task of environmental economists therefore is to transform ecological assets into marketable goods. For example, if clean air is not a marketable good with a price, then the market places no value on it. Thus when an industrial plant emits air pollution, it simply externalizes the cost (which shows up in premature deaths, damage to ecosystems, deterioration of environmental amenities, etc.) to society, while the environmental damage is not internalized within the market or on the balance sheet of the firm. The answer to this, from the standpoint of neoclassical environmental economics, is to create markets in clean air, thereby internalizing such external costs within the market. The overall logic is one of bringing the earth within the balance sheet.[4]

Costing the Earth

Since the environment (that is, the biosphere) is not a commodity, however, and is not reproduced according to the rules of the market, what means are to be adopted in order to internalize the environment within the market system? It is here that most of the attention of environmental economics is directed. Neoclassical environmental economists essentially rely on a three-stage process. First, they break the environment down into specific goods and services, separated out from the biosphere and even from particular ecosystems, in such a way as to make them into commodities (to a degree), for example, the timber available in a particular forest, the water quality in a given river, the species in a particular wildlife reserve, or the maintenance of a certain global temperature over a number of decades. Then these goods and services are given an imputed price through the construction of supply and demand curves, presumably allowing economists to ascertain the optimal level of environmental protection.[5] Finally, various market mechanisms and policy instruments are devised in order to either change prices in existing markets or to create new markets so as to achieve the desired level of environmental protection.

A great deal of attention is given in this process to the construction of demand curves for environmental goods and services. (The task of constructing supply curves, associated with the costs of environmental protection, is generally considered—perhaps mistakenly—to pose fewer difficulties than the demand curve.) Demand curves are constructed by determining the willingness to pay of consumers. However, since actual markets for environmental goods and services do not for the most part exist—that is, these products are not actually bought—the willingness to pay on the part of consumers is imputed in a couple of ways.

The first of these methods is known as hedonic pricing. In this approach consumer preferences are supposedly revealed through the demand for goods and services that are closely associated with a given environmental product. Such closely associated goods and services, existing within actual markets, are seen as in some way standing in for the environmental product in question, or else offering the basis for comparisons from which calculations regarding the willingness to pay for a given environmental product can be derived. For example, the willingness of consumers to pay for a quiet neighborhood is calculated by comparing the market price of homes near an airport with similar homes in a quieter locale. Or the willingness of people to protect a recreation site can be imputed on the basis of their willingness to pay transportation costs to visit the site.

An example of hedonic pricing in the United States occurs in government attempts to value the environmental assets that would be lost—say in the construction of a dam—by letting the amount that sportsmen (and sportswomen) pay on average in their pursuit of fish and game stand in for the value of these species, which is taken as an indication of the value of a given local ecosystem. This is accomplished through the use of a concept known as wildlife fish user days (WFUD), representing the amount of money that an average individual sportsperson could be expected to spend in 12 hours in pursuit of various forms of wildlife. Water fowl were valued in the early 1980s at $19–$32 per WFUD, elk at $16–$25 per WFUD, and fish at $14–$21 per WFUD. By this means the market utility associated with the pursuit of fish and game (representing the demand for environmental protection in the area to be flooded) could be compared, within the context of a broader cost-benefit analysis, with the market utility to be derived from (and the willingness of consumers to pay) for a new dam, or some other development project.[6]

Such bottom-line thinking recognizes no boundaries outside of the accounting ledger. A closely related form of cost-benefit analysis has been applied to human beings, in assessing risk within the occupational environment. Under the Reagan administration, the Office of Management and Budget (OMB) attempted to promote calculations of the *dollar value of a human life* based on the *wage premiums* that workers required in order to accept a higher risk of early death. On this basis a number of academic studies concluded that the value of a worker's life in the United States (in the early 1980s) was worth between $500,000 and $2 million (far less than the annual salary of many corporate chief executive officers). The OMB then used these results to argue that certain forms of pollution abatement were cost-effective, while others were not, in accordance with President Reagan's Executive Order No. 12291 that regulatory measures should "be chosen to maximize the net benefit to society."[7]

The second major method of determining consumer preferences is what is known as the contingent valuation method. Here hypothetical markets are constructed and consumers are asked to indicate their preferences through surveys. In such surveys representative samples of the population are asked what they would pay for a given level of protection for a given environmental commodity, and at the same time what they would have to be given in compensation for losing it. Ideally, such surveys should cover a large number of levels of protection, generating a whole range of values from consumers. This seldom proves practicable, however, and more often the surveys cover only a few levels of protection—for example, making a river fishable, swimmable, and drinkable. Using the responses of individuals in these surveys, economists aggregate the results across the entire population in order to construct demand curves for the hypothetical environmental commodities.

Having determined the most appropriate (most cost efficient) level of environmental protection through hedonic pricing and contingent valuation, neoclassical economists move on to the problem of how to alter existing markets or to create new ones in order to achieve the optimal level of protection.[8] Much of environmental economics thus aims at the creation of markets to solve problems of pollution and environmental degradation. Essentially, there are two market-oriented techniques used.[9] One is the fairly straightforward imposition of taxes or subsidies that will increase the costs of inflicting environmental damage and the benefits of environmental improvements. The other technique is to use the state to create new markets, which then operate on their own. One example of this is the charging of entrance fees to parks, so as to restrict admission in accordance with ability to pay. Another is changing property rights, such as the creation of exclusive economic zones in coastal waters. Particularly popular among neoclassical environmental economists and policy makers is the use of the state to establish market-based incentives such as tradable pollution permits. This allows pollution up to a certain overall level while making it possible for firms that are more efficient in reducing pollution to benefit through the sale of these permits—thus forcing those firms that continue to pollute excessively to pay for their pollution.

The entire neoclassical view, it should be clear beyond any doubt at this point, rests on turning the environment into a set of commodities.[10] Further, the goal is quite explicitly one of overcoming the so-called market failures of the environment by constructing replacement markets for environmental products. If environmental degradation and pollution are evident, the economist reasons, it must be because the environment has not been fully incorporated within the market economy, and does not operate according to the laws of economic supply and demand. Yet, the faulty character of neoclassical environmental economics becomes evident when one realizes that this entire methodology is based on the utopian myth that the environment can and should become a part of a self-regulating market system.

Contradictions of Economic Reductionism

Nature, however, is not a commodity produced to be sold on the market according to economic laws of supply and demand. Nor is it a market organized according to laws of individual consumer preferences. It is not even privately owned for the most part. The allocation and distribution of environmental goods is subject to state regulation.

The environment can be rationally considered a "condition of production" for the economy. However, it cannot be fully incorporated into the circular flow of a commodity economy. There are ethical reasons why we may choose to preserve crucial parts of nature from the forces of the market. Moreover, any attempt to allow the "tyranny of the bottom line" guide our relation to nature in its entirety would be disastrous.[11]

The conditions of environmental reproduction (that is, ecological sustainability) can be undermined not only through the economy failing to take environmental costs into account (the externalization of costs to the environment), as is commonly supposed, but also by the attempted incorporation of the environment into the economy—the commodification of nature. The reason for this is that the underlying problem can be traced neither to the nature of the environment itself nor to market failures (imperfections in the workings of the market system), but rather arises from the fundamental nature of the socioeconomic system in which we live.

"What we call land," Karl Polanyi wrote in *The Great Transformation,*

is an element of nature inexplicably interwoven with man's institutions. To isolate it and form a market out of it was perhaps the weirdest of all undertakings of our ancestors. . . . The economic function is but one of many vital functions of land. It invests man's life with stability; it is the site of his habitation; it is a condition of his physical safety; it is the landscape and the seasons. We might as well imagine his being born without hands and feet as carrying on his life without land. And yet to separate land from man and organize society in such a way as to satisfy the requirements of a real estate market was a vital part of the utopian concept of a market economy.[12]

The "weird" nature of such a reductionist approach to nature, arising out of an attempt to construct not only all of society but also the entire ecology of humankind (and indeed ecological relations in general along market-commodity lines), has its concrete manifestation in three interwoven contradictions. The first is the radical break with all previous human history necessitated by the reduction of the human relation to nature to a set of market-based utilities, rooted in the egoistic preferences of individuals. "For the first time," Marx wrote of capitalist society, "nature becomes purely an object for humankind, purely a matter of utility; ceases to be recognized as a power for itself; and the theoretical discovery of its autonomous laws appears merely as a ruse so as to subject it under human needs, whether as an object of consumption or as a means of production."[13] By reducing the human relation to nature to purely possessive, individual terms, capitalism thus represents (in spite of all of its technological progress) not so much a fuller development of human needs and powers in relation to the powers of nature, as the alienation of nature from society in order to develop a one-sided, egoistic relation to the world.

The second contradiction of economic reductionism when applied to nature is associated with the radical displacement of the very idea of value or worth, resulting from the domination of market values over everything else. This alienation of nature was highlighted by Kant's classic distinction between market price and intrinsic value: "That which is related to general human inclinations and needs has a market price. But that which constitutes the condition under which alone something can be an end in itself does not have mere relative worth (price) but an intrinsic worth (dignity)."[14]

It is this widespread humanistic sense of systems of intrinsic value that are not reducible to mere market values and cannot be included within a cost-benefit analysis that so often frustrates the attempts of economists to carry out contingent value analyses among the general public. If asked whether the market economy should place a value on all of nature, most people would probably say yes, but this really tells us nothing, given that the concept of value in ordinary parlance may often mean something like dignity rather than price. Time and again, when asked to set a price on particular environmental goods such as air quality or a picturesque landscape, large numbers of people will simply refuse—sometimes with the suspicion that such exercises point not to preservation but to something more like a protection racket.[15]

For many, probably a majority of people—even in our self-centered, acquisitive society—nature is not something to be broken into pieces and then inserted into a system of relative prices. Viewing nature in terms of individual consumer preferences rather than convictions, duties, aesthetic judgments, etc. is for most people a kind of "category mistake."[16] As E. F. Schumacher noted in his critique of the application of cost-benefit analysis to the environment in his *Small is Beautiful,* the attempt "to measure the immeasurable is absurd and constitutes [on the part of the economist] but an elaborate method of moving from preconceived notions to foregone conclusions; all one has to do to obtain the desired results [the reduction of parts of the environment to commodity values] is to impute suitable values to the immeasurable costs and benefits." What is worst about this undertaking "is the pretence that everything has a price or, in other words, that money is the highest of all values."[17]

The third contradiction of economic reductionism when applied to the environment can be seen in the material consequences, not merely those of a moral nature. Although the internalization of the environment within the economy, by providing commodity prices for everything in nature, and establishing markets (often by artificial means) to solve all problems of pollution, resource exhaustion, etc., is often presented as the way out of our ecological problems, a good case can be made that such solutions, while sometimes attenuating the problems in the short term, only accentuate the contradictions overall, undermining both the conditions of life and the conditions of production. The reason for this is the sheer dynamism of the capitalist commodity economy, which by its very nature accepts no barriers outside of itself, and seeks constantly to increase its sphere of influence without regard to the effects of this on the biosphere. It is not so much the failure to internalize large parts of nature into the economy that is the source of environmental problems, but rather that more and more of nature is reduced to mere cash nexus and is not treated in accordance with broader, more ecological principles.

For a neoclassical economist, songbird species are facing extinction because their relative prices are too low (that is, they are outside the market). The "natural" solution from this standpoint then is to find a way of bidding up the price of songbirds by creating markets for them. However, finding a way of assigning a higher relative price to songbirds is unlikely to do much good as long as the primary reason for their approaching extinction is expansion of the entire system of contemporary agribusiness, with its disastrous (and frequently poisonous) effect on the habitat on which these birds depend.

Similar issues arise in the case of forest ecosystems, although in this situation it is not a question of being outside the market or lacking a price tag. Forests have long been managed on market-based principles. The result in most cases has been the loss of forests, since the market sees forests, not as ecosystems, but as consisting of so many million or billion board feet of standing timber. According to the rules of accumulation under a system of market values, a relatively untouched, pristine forest (i.e., an intact forest ecosystem supporting diverse species) is a "gift of nature" not yet in a fully managed state, containing trees that may be hundreds of years old and that are no longer growing at a rate that is justified according to the current rate of interest. Such unproductive assets therefore need to be harvested as quickly as possible and replaced by an industrial tree plantation consisting of a single species of trees of uniform age, grown with the help of a massive infusion of industrial chemicals, and harvested and turned into commodities within a few decades. Such monocultures no longer support a diverse range of plant and animal species. In effect, an extreme division (and simplification) of nature has occurred because it has been turned into a commodity. Forest ecosystems are threatened therefore, not by the failure to incorporate them into the market system, but rather by the "natural" operations of the commodity system itself, and by the extreme narrowness of its objectives.[18]

From an ecological standpoint, insofar as the diversity of life is an objective, the market is extremely inefficient compared with nature itself. Encountering a tropical forest for the first time, the great nineteenth-century naturalist Alexander von Humboldt remarked that the very density of the forest "enlarged the domains of organic nature."[19] This principle, central to Darwin's evolutionary theory, came to be known as the law of divergence—the more diverse forms of life in a given area (the more ecological niches seized upon), the more that area will support. Yet turning forests into commodities has led to their degradation (i.e., extreme simplification), thereby *diminishing* rather than enlarging the domains of organic nature in this sense.

The Ideology of Natural Capital

Recently, it has become popular among environmental economists—who are well aware of the destructive impact that the commodity economy has had on nature—to argue that the internalization of environmental costs has to be seen in a broader way by recognizing that all of nature and its various components are essentially "natural capital."[20] Such economists present the environmental crisis not so much as a failure of the market as a failure of our accounting system, which does not recognize that capital already includes all of existence.

Green entrepreneur Paul Hawken has popularized the view that true capitalism, as opposed to capitalism as we have known it thus far, would take natural capital into account, and by doing so save the environment. The system of market pricing has failed us with respect to the environment, he writes, "for the most simple and frustrating of reasons: bad accounting. Natural capital has never been placed on the balance sheets of companies, countries or the world. Paraphrasing G. K. Chesterton, it could be fairly said that capitalism might be a good idea except that we have never tried it yet. And try it we must and will, for capitalism cannot be fully attained or practiced until, as any accounting statement will tell us, we have an accurate balance statement."[21]

Ecological-socialist economist Martin O'Connor has referred to this theoretical tendency critically as the attempted "capitalization of nature," meaning "the *representation* of the biophysical milieu (nature) and of nonindustrialized economies and the human domestic sphere (human nature) as reservoirs of 'capital,' and the codification of these stocks as property tradable 'in the marketplace.'"[22]

The proper domain of capital is thus magically enlarged by a mere change of terminology. Formerly, all of nature was treated as a "gift" to capital and as an external and exploitable domain. Now it is increasingly "redefined as itself a stock of capital." Correspondingly, the nature of capitalism is seen as changing "from accumulation and growth feeding on an external domain to ostensible self-management and conservation of the *system of capitalist nature* closing back on itself."[23] The irony here is that capitalism, in typical fashion, sees any crisis as emanating from barriers to the expansion of capital rather than the expansion of capital itself. The solution is to increase the domain of capital, recognizing that nature too is properly part of the rational system of commodity exchange.

Just as Weber admitted in his *General Economic History* that historically capitalism had been based on rapacious colonialism, but went on to deny that this had anything to do with modern rational capitalism, which no longer relied on force or unequal exchange, so the contemporary environmental economists argue that capitalism historically relied on a rapacious relation to nature, but that a modern rational capitalism— capitalism worthy of the name—is destined to bring all of nature within its balance sheet.[24] "After all," as Martin O'Connor observes, "if capital is nature and nature is capital, the terms become virtually interchangeable; one is in every respect concerned with *the reproduction of capital, which is synonymous with saving nature.* The planet as a whole is our capital, *which must be sustainably managed.*"[25]

Yet with all the rhetoric of the valuation of natural capital, the actual operation of the system has not materially changed, and can't be expected to change. The rhetoric of nature and the planet as capital thus serves mainly to obscure the reality of the extreme exploitation of nature for the sake of commodity exchange. Moreover, the principal result of the incorporation of such natural capital into the capitalist system of commodity production—even if carried out—will be the further subordination of nature to the needs of commodity exchange. There will be no actual net accumulation of natural capital; rather, nature will increasingly be converted into money or abstract exchange, subject to the vicissitudes of Wall Street. "The commodity," as eco-Marxist economist Elmar Altvater has observed, "is narcissistic; it sees only itself reflected in gold."[26]

Today parts of the redwood forest of northern California that are under private management (Pacific Lumber Corp.) are being removed because trees that are centuries old are considered nonproducing assets, and the rules of the market (and Wall Street) demand that they be liquidated and replaced by younger, faster-growing trees, which can be placed in a "fully managed" condition. The tragic fate of these forests—as noted earlier—is not due to their exclusion from the capitalist balance sheet, but rather to their inclusion. The market has no internal mechanism that recognizes that the results of such decisions are irreversible within the normal human time span (it would take many generations to repair the damage, even if the system would allow such an enormously costly—in terms of market exchange—process of restoration).

The new hegemonic vision of environmental economics thus seeks to extend the domain of capital to all of nature as the means of preserving the latter. "In what we might call the *ecological phase of capital,*" Martin O'Connor critically observes, "the relevant image is no longer of man acting on nature to 'produce' value, henceforth appropriated by the capitalist class. Rather, the image is of nature (and human nature) codified as *capital incarnate,* regenerating itself through time by controlled regimes of investment around the globe, all integrated in a 'rational calculus of production and exchange,' through the miracle of a price system extending across space and time. This is nature conceived in the image of capital."[27]

Accumulation and the Environment

The principal characteristic of capitalism, which this whole market-utopian notion of the capitalization of nature ignores, is that it is a system of self-expanding value in which accumulation of economic surplus—rooted in exploitation and given the force of law by competition—must occur on an ever-larger scale. At the same time, this represents a narrow form of expansion that dissolves all qualitative relations into quantitative ones, and specifically in monetary or exchange value terms. The general formula for capital (generalized commodity production), as Marx explained, is one of M-C-M', whereby money is exchanged for a commodity (or the means of producing a commodity), which is then sold again for money, but with a profit. This expresses capitalism's overriding goal: the expansion of money values (M'), not the satisfaction of human needs. The production of commodities (C) is simply the means to that end.

The ceaseless expansion that characterizes such a system is obvious. As the great conservative economist Joseph Schumpeter remarked, "capitalism is a process, stationary capitalism would be a *contradictio in adjecto.*"[28] Economists, even environmental economists, rarely deal with the question of the effect that an increasing economic scale resulting from ceaseless economic growth will have on the environment. Most economists treat the economy as if it were suspended in space, not as a subsystem within a larger biosphere. Moreover, many economists who recognize the importance of natural capital nevertheless adopt what is known as the "weak sustainability hypothesis." According to this hypothesis, increases in the value of human capital fully compensate for any losses in natural capital, such as forests, fish stocks, and petroleum reserves.

Some ecological economists, however, have countered with what is known as the "strong sustainability hypothesis," according to which human-made capital cannot always substitute for natural capital, since there is such a thing as critical natural capital, that is, natural capital necessary for the maintenance of the biosphere. Tropical forests, for example, are home to about half the world's species and are critical in regulating the planet's climate. Once this is admitted though, the dream of reducing all of nature to natural capital to be incorporated within the market fades quite quickly. The self-regulating market system has no way of valuing nature on such a scale. Moreover, there is an inherent conflict between the maintenance of ecosystems and the biosphere and the kind of rapid, unbounded economic growth that capitalism represents.[29]

Indeed, sustainable development envisioned as the "pricing of the planet" (to refer to the title of one recent book) is little more than economic imperialism vis à vis nature.[30] It tends to avoid two core issues: whether all environmental costs can actually be internalized within the context of a profit-making economy, and how the internalization of such costs can account for the effects of increasing economic scale within a limited biosphere. The difficulty of internalizing all external costs becomes obvious when one considers what it would take to internalize the costs to society and the planet of the automobile-petroleum complex alone, which is degrading our cities, the planetary atmosphere, and human life itself. Indeed, as the great ecological economist K. William Kapp once remarked, "Capitalism must be regarded as an economy of unpaid costs."[31] The full internalization of social and environmental costs within the structure of the private market is unthinkable.

To be sure, some advocates of natural capitalism, like Paul Hawken, contend that economic growth itself on whatever scale raises no insurmountable obstacle to the environment. This, however, has to take the form of unconventional growth characterized by the dematerialization of the economy—reductions in the throughput of raw materials and energy per unit of output. Hawken points to the possibility of a 200-mile-per-gallon car and what he calls the "magic carpet" of recycling.[32] However, thinking that such technological wonders can resolve the problem not only goes against the basic laws of thermodynamics (specifically the entropy law, which tells us that nothing comes from nothing) but also defies all that we know about the workings of capitalism itself, where technological change is subordinated to market imperatives. The biggest obstacle to automobiles with greater gas efficiency is posed by the whole automobile-petroleum complex, i.e., the most powerful corporations in the world. At best, as Altvater has noted, "The economic 'internalization' of economic effects is only a stop-gap; it would not, in any imaginable situation, compensate for the way in which natural conditions are altered through the 'throughput' of materials and energy in production, consumption, and even distribution."[33]

The Ecological Blinders of Neoclassical Economics

The case of global warming illustrates well the conservative nature of economics when confronted with existing and impending environmental catastrophes. In an attempt to guess the costs of global warming over the next century, the prestigious neoclassical economist William Nordhaus, writing for *Science* magazine in 1992, suggested that the costs would be largely in the agricultural realm, the main market sector to be affected, and came up with a figure of a 1 percent loss in gross national product (GNP).[34] When scientists criticized this estimate as being hopelessly naive, Nordhaus admitted in a later article that his guess had not taken into account the effects that heating of the earth would have on nonmarket sectors, i.e., the value of species driven into extinction and wetlands lost with rising sea levels, the costs associated with the creation of environmental refugees, etc. Since these nonmarket costs are hard to measure, Nordhaus had resorted to a limited sampling of the opinions of economists, atmospheric scientists, and ecologists (so-called expert opinion) on what the costs would be.[35] What was revealed, not surprisingly, was an immense cultural divide. As the atmospheric scientist Stephen Schneider, one of Nordhaus' critics, summarized the results:

The most striking difference in the [1994 Nordhaus] study was that almost all the conventional economists considered even a radical scenario in which a 6°C warming would unfold by the end of the next century (a scenario I would label as catastrophic, but improbable—maybe only a 10 percent chance of occurring) as not very catastrophic economically. Most conventional economists still thought even this gargantuan climatic change—equivalent to the scale of change from an ice age to an interglacial epoch in a hundred years, rather than thousands of years—would have only a few percent impact on the world economy. In essence, they accept the paradigm that society is almost independent of nature. In their opinion, most natural services associated with current climate can be substituted with relatively little harm to the economy.

On the other hand, the group Nordhaus labeled as natural scientists thought the damage to the economy from the severe climate change scenario would range from a loss in GNP of several percent to 100 percent; the latter expert assigned a 10 percent chance to the virtual destruction of civilization! Nordhaus suggested that the ones who know the most about the economy are optimistic. Schneider countered with the obvious retort that the ones who know the most about the environment are worried.[36]

On the basis of the estimates of the economic costs of global warming projected by Nordhaus and other economists, such as William Cline, some economists, most notably Lawrence Summers, formerly chief economist of the World Bank and assistant secretary of agriculture for international economic affairs in the Clinton administration, have argued that there are no strong economic reasons for moving fast on global warming. Indeed, adopting the weak sustainability hypothesis, Summers contended that "we can help our descendants as much by improving infrastructure as by preserving rain forests."[37]

This failure of economists to understand that human society and the human economy exist within a larger biosphere and that undermining the conditions of life is bound to undermine the conditions of production takes us to the heart of the failure of both neoclassical economics and the self-regulating market system itself. Nature is not a commodity and any attempt to treat it as such and to make it subject to the laws of the self-regulating market is therefore irrational, leading to the overexploitation of the biosphere by failing to reproduce the conditions necessary for its continued existence.

As the scale of the world commodity economy has grown, so have the number and scale of our ecological problems: global warming, destruction of the ozone layer, extinction of species, loss of genetic diversity, the annihilation of tropical rain forests, desertification, the spread of toxic wastes, pollution of oceans, the decline in environmental health, etc. Although these problems are in many ways discrete, they are also interrelated and have their source in the effects of the commodity economy on nature—whether by the externalization of costs or through the internalization of nature into an economy geared to the unlimited growth of capital.

Those arguing from an economic point of view sometimes say that as ecological resources become scarce, the economy will respond by moving toward preservation. Yet such a smooth functional relationship does not exist. As the radical Green Rudolf Bahro wrote, "The rising cost of land has never been able to halt the building up and concretizing over of the landscape."[38] Nor is it possible to solve the problem by applying what is known as "the polluter pays principle," whereby costs are inflicted on the individual polluters. Such views deny the systematic and interrelated nature of the problem: Entire industrial complexes are involved, and ultimately it becomes a question of the expansion of the market itself. Ecological reforms within the system, like all other reforms, are limited because the moment they begin to address the fundamental nature of the system itself, they are quickly curtailed by the vested interests.

Beyond the Bottom Line

Ultimately the defense of the environment therefore requires a break with the tyranny of the bottom line and a long revolution (it is hoped, not too long, given the acceleration of history associated with ecological change) in which other, more diverse values not connected to the bottom line of the money-driven economy have a chance of coming to the fore. What is needed is a system of production organized democratically in accordance with the needs of the direct producers and reflecting an emphasis on the fulfillment of the totality of human needs (extending beyond the Hobbesian individual).[39] These have to be understood as connected to the sustainability of nature, i.e., the conditions of life as we know them. Production can be said to be nonalienating only if it promotes the welfare of every individual as the way of promoting the welfare of all, and only if it fulfills the human need for a sustainable, and in that sense nonexploitative, relation to nature.

Since environmental costs under capitalism tend to be externalized while the benefits of avaricious disregard of environmental necessity feed the wealth of the few, environmental depredations lead to struggles for environmental justice. The struggle for material welfare among the great mass of the population, which was once understood mainly in economic terms, is increasingly taking on a wider, more holistic environmental context. Hence, it is the struggle for environmental justice—the struggle over the interrelationship of race, class, gender, and imperial oppression and the depredation of the environment—that is likely to be the defining feature of the twenty-first century. The universalization of a capitalism that knows no bounds is unifying all that seek to exist in defiance of the system. Historic struggles for social justice are becoming united, as never before, with struggles for the preservation of the earth. The solution to the environmental problem, our own struggles will teach us, lies beyond the bottom line. It is here that the main resources for hope in the twenty-first century are to be found.

Notes

1. Within the broader field of environmental economics, it has become common to distinguish between those who are "environmental economists" proper (that is, neoclassical environmental economists) and those who adhere to "ecological economics." The latter are distinguished by their application of thermodynamics (the entropy law) to economics in the tradition of Nicholas Georgescu-Roegen, by their greater emphasis on limits to growth (or the problem of economic scale in relation to the environment), and by their insistence that fundamental transformations in values and institutions are necessary to cope with the deepening ecological crisis. Clearly, most economists working on environmental issues at present are working within the tradition of neoclassical environmental economics, the goal of which is to make current institutions work better, without in anyway questioning the fundamental values of the self-regulating market system. Hence, it is with this dominant tradition, and not with "ecological economists," that this essay is mainly concerned.

2. Michael Jacobs, "The Limits to Neoclassicism: Towards an Institutional Environmental Economics," in *Social Theory and the Environment,* Michael Redclift and Ted Benton, eds. (New York: Routledge, 1994, p. 69); George Eads and Michael Fix, *Relief or Reform?: Reagan's Regulatory Dilemma* (Washington, D.C.: Urban Institute, 1984, p. 14). The first part of this chapter draws heavily on Jacobs' admirable critique of neoclassical environmental economics.

3. This is stated explicitly by the influential British environmental economist D. W. Pearce, who writes in the opening sentence of his book *Environmental Economics,* "In approaching the subject matter of environmental economics it is important to understand that, with some exceptions, economists have regarded environmental degradation as a particular instance of 'market failure.'" D. W. Pearce, *Environmental Economics* (New York: Longman, 1976, p. 1).

4. See R. Kerry Turner, David Pearce, and Ian Bateman, *Environmental Economics* (Baltimore, Md.: Johns Hopkins Univ. Press, 1993, pp. 75–77).

5. See Jacobs, "Limits of Neoclassicism," pp. 70–71.

6. Marilyn Waring, *If Women Counted* (New York: HarperCollins, 1988, p. 267); *National Wildlife* (April/May 1986):12.

7. Barry Commoner, *Making Peace with the Planet* (New York: New Press, 1992, pp. 64–66).

8. It should be noted, however, that not all neoclassical environmental economists are concerned with such issues as hedonic pricing and contingent evaluation. It is possible to skip the stage of construction of demand curves that indicate consumer preferences and to derive the initial criteria for the level of protection desired through political or scientific means. Logically, all that is required at this stage is the assumption that markets constitute the most efficient way in which to achieve a given level of environmental protection—regardless of how those preferences were derived in the first place.

9. These techniques exclude more direct state regulation (or command and control) divorced from the market-oriented approaches preferred by neoclassical environmental economists.

10. Jacobs, "Limits of Neoclassicism," p. 74.

11. The phrase "tyranny of the bottom line" is taken from Ralph W. Estes, *Tyranny of the Bottom Line: Why Corporations Make Good People Do Bad Things* (San Francisco: Berrett-Koehler, 1996).

12. Karl Polayni, *The Great Transformation* (Boston: Beacon Press, 1944, p. 178).

13. Karl Marx, *Grundrisse* (New York: Vintage, 1973, pp. 409–10). In referring to the "ruse" whereby the systematic understanding of nature's laws is viewed simply as a means of subjecting nature to human ends, Marx is clearly referring to the well-known maxim of Bacon: "Nature is only overcome by obeying her." Francis Bacon, *Novum Organum* (Chicago: Open Court, 1994, pp. 29, 43).

14. Immanuel Kant, *Foundations of the Metaphysics of Morals* (Upper Saddle River, N.J.: Prentice-Hall, 1995, pp. 51–52).

15. Mark Sagoff, *The Economy of the Earth* (New York: Cambridge Univ. Press, 1988, p. 88).

16. Ibid., pp. 92–94.

17. E. F. Schumacher, *Small Is Beautiful* (New York: Harper & Row, 1973, p. 46).

18. The consequences of this process can be seen in the destruction of the old-growth forest of the Pacific Northwest. See John Bellamy Foster, "The Limits of Environmentalism Without Class: Lessons from the Ancient Forest Struggle in the Pacific Northwest," in *The Struggle for Ecological Democracy*, Daniel Faber, ed. (New York: Guilford Press, 1998, pp. 188–217).

19. Alexander von Humboldt, quoted in Loren Eisley, *Darwin's Century* (New York: Anchor Books, 1961, p. 183).

20. See, for example, Thomas Prugh, *Natural Capital and Human Economic Survival* (Solomons, Md.: International Society for Ecological Economics, 1995). The concept of "natural capital" itself dates back to the ecological economics of the 1850s in the U.S., in the work of such thinkers as George Waring and Henry Carey. See George E. Waring, Jr., "Agricultural Features of the Census of the United States for 1850," *Organization & Environment* 12(3):305–06.

21. Hawken, foreword in Prugh, *Natural Capital,* p. xiv.

22. Martin O'Connor, "On the Misadventures of Capitalist Nature," in *Is Capitalism Sustainable?* Martin O'Connor, ed. (New York: Guilford Press, 1994, p. 126).

23. Ibid.

24. Max Weber, *General Economic History* (New Brunswick, N.J.: Transaction, 1981, p. 300).

25. O'Connor, "On the Misadventures," pp. 132–33.

26. Elmar Altvater, *The Future of the Market* (New York: Verso, 1993, p. 184).

27. Ibid., p. 131.

28. Joseph Schumpeter, *Essays* (Reading, Mass.: Addison-Wesley, 1951, p. 293).

29. On the weak and strong sustainability hypotheses, see Turner et al., *Environmental Economics,* pp. 31, 54–56.

30. Peter H. May and Ronaldo Seroa Motta (eds.), *Pricing the Planet* (New York: Columbia Univ. Press, 1996).

31. K. William Kapp, *The Social Costs of Private Enterprise* (New York: Shocken Books, 1971, p. 231).

32. See Paul Hawken, "Natural Capitalism," *Mother Jones Magazine* (April 1997): 40–53, 59–62; John Bellamy Foster, "Natural Capitalism?" *Dollars & Sense* (May–June 1997):9.

33. Altvater, *Future of the Market,* p. 186.

34. William Nordhaus, "An Optimal Transition Path for Controlling Greenhouse Gases," *Science* **258** (November 20, 1992):1316.

35. William Nordhaus, "Expert Opinion on Climate Change," *American Scientist,* 82(1) (January/February 1994):45–51.

36. Schneider, *Laboratory Earth* (New York: Basic Books, 1997, pp. 133–34).

37. Lawrence Summers, *The Economist* (May 30, 1992): 65.

38. Rudolf Bahro, *Avoiding Social and Ecological Disaster* (Bath: Gateway Books, 1994, p. 50).

39. For a discussion of the environmentalist notion of the "acceleration of history" (promoted in particular by the Worldwatch Institute), see John Bellamy Foster, *The Vulnerable Planet* (New York: Monthly Review Press, 1999), pp. 143–49.

II
Shaping Consciousness

9

Silencing Spring: Corporate Propaganda and the Takeover of the Environmental Movement

Sheldon Rampton and John Stauber

At the time of the American Revolution, neighbors and townfolk participated in grassroots decision-making, sharing their opinions with each other directly and through a rich array of communication forums that included public meetings and political debates. While wealth governed politics, it was still a far cry from today's level of influence peddling. Today's political environment has been profoundly shaped by a public relations industry that works to obfuscate reality and manipulate the political process. As citizens increasingly remove themselves in disgust from what they correctly perceive as a degraded political process, the PR industry takes their place, turning the definition of grassroots politics upside down to serve the interests of their corporate clients.

This chapter explores the daily tactics of the PR industry, with particular emphasis on how it co-opts and disorganizes the environmental movement and shifts attention to the individual as the source of the environmental crisis. If corporations are not despoiling our natural environment, then *who is to blame?* According to corporate-sponsored propaganda campaigns, you are. In place of systemic analysis and systemic solutions, corporate PR offers individualistic and deeply hypocritical analysis in which "all of us" are to blame for our collective "irresponsibility."

Propaganda, as Australian scholar Alex Carey observes, flourishes most commonly in democratic societies. Dictatorships tend to rely on comparatively crude and unpersuasive propaganda because they do not usually need to brainwash their subjects—they simply kill or jail the people who oppose them. Indeed, the United States has been noteworthy both for its level of individual freedom and for the sophistication, variety, and sheer scale of its propaganda industries.[1] Advertising is only the visible tip of the iceberg. Invisible propaganda, in the form of public relations, is a major industry. Unlike advertising, which openly broadcasts the identity of its sponsors, public relations intentionally conceals much of its workings from the public. "The best PR ends up looking like news," brags one public relations executive. "You never know when a PR agency is being effective; you'll just find your views slowly shifting."[2]

The public relations industry, which did not exist before the twentieth century, has grown into an institution so pervasive that part of its invisibility stems from its being everywhere, from T-shirts bearing cigarette brand names to "citizens' groups" which turn out upon close examination to be organized, led, and funded by corporate interests promoting private, for-profit agendas. No one knows exactly how much money corporations spend annually in the United States on public relations, but $10 billion is considered a conservative estimate. PR professionals use sophisticated psychology, opinion polling, and complex computer databases so refined that they can pinpoint the prevailing "psychographics" of individual city neighborhoods.

Press agents used to rely on news releases and publicity stunts to attract attention. Today, raw money enables the PR industry to mobilize 800 numbers and telemarketing, advanced databases, satellite feeds, computer bulletin boards, and simultaneous multilocation fax transmission to communicate its message. Ordinary citizens have the right to organize for social justice and freedom from toxins, but they cannot afford multi-million-dollar campaigns.

The public relations industry emerged in response to corporate and government fears of the "chaos" associated with democracy, a fear expressed in French social philosopher Gustave Le Bon's warning that "the divine right of the masses is about to replace the divine right of kings." Prior to the twentieth century, corporations could afford to espouse a "public be damned" attitude, but the rise of the Progressive movement and journalist muckrakers like Upton Sinclair brought new recognition that uncontrolled public opinion could threaten corporate profits and even the capitalist system itself.

The self-proclaimed "father" of public relations, Edward Bernays, drew upon psychological theory as he built a profession based upon the belief that human beings are irrational and unconscious of their own real motives. In books such as *Engineering Consent* and *Propaganda,* Bernays defined public relations as an "applied social science" that society's masters could use to manage the human herd. "If we understand the mechanism and motives of the group mind," he argued, it would be possible to "control and regiment the masses according to our will without their knowing it. . . . Theory and practice have combined with sufficient success to permit us to know that in certain cases we can effect some change in public opinion with a fair degree of accuracy. . . ."[3]

Bernays perfected an approach that PR professionals have come to call "the third party technique." Merrill Rose, executive vice-president of the public relations firm Porter/Novelli, explains the technique succinctly: "Put your words in someone else's mouth."[4] This deliberately deceptive strategy offers several advantages from a propagandist's perspective:

1. *It offers camouflage,* helping to hide the profit-driven self-interest that lurks behind the message, subverting the public's ability to recognize and analyze its propagandistic nature.

2. *It encourages conformity to a corporate agenda while feigning independence.*

3. *It deliberately weds that corporate agenda to a popular, even progressive cause,* such as the environment. This element of the "third party technique" recurs so frequently that it can be rightly described as a central organizing strategy of the public relations industry.

4. *It drives out factual discourse and replaces it with emotional imagery,* in which the identity of the messenger becomes symbolic, at an unconscious level, of the desired message.

The strategy of "putting your words in someone else's mouth" takes many forms. In dealing with issues affecting the environment and human health, a few of these strategies deserve particular attention.

Experts for Hire

Public relations firms have discovered the value of sponsoring "experts" to serve as third-party advocates for their sponsors' views. According to a survey commissioned by the Porter/Novelli PR firm, 89 percent of respondents consider "independent experts" a "very or somewhat believable source of information during a time of 'corporate crisis.' "[5] Experts are used both to tout the virtues of a company's products and to downplay their hazards.

A host of techniques exist for manipulating research protocols to produce studies whose conclusions fit the sponsor's predetermined interests. These techniques include adjusting the time of the study, subtle manipulations of control groups or dosage levels, and subjective interpretations of complex data. Often such methods stop short of outright fraud, but lead to predictable results. "Usually associations that sponsor research have a fairly good idea what the outcome will be, or they won't fund it," admits Joseph Hotchkiss of Cornell University.[6]

In *Tainted Truth: The Manipulation of Fact in America,* author Cynthia Crossen warns that "Hundreds of private firms also do research for hire. Whatever their predelictions—or financing—researchers often affix adjectives like 'independent' and 'nonprofit' to their neutral names. The Health Effects Institute/Asbestos Research, which describes itself as a private nonprofit Cambridge-based research organization, is partly funded by the federal government and the rest by industry groups with a financial stake in the outcome of the institute's research."[7]

In recent years, declining public funding for research has enabled corporate sponsors to form partnerships even with prestigious institutions such as major universities, with similar predetermined outcomes. *The New England Journal of Medicine (NEJM)*, often described as the world's most prestigious medical journal, has been featured repeatedly in controversies regarding hidden economic interests that shape its content and conclusions. In 1986, for example, *NEJM* published one study and rejected another that reached opposite conclusions about the antibiotic amoxicillin, even though both studies were based on the same data. Scientists involved with the first favorable study had received $1.6 million in grants from the drug manufacturer, while the author of the critical study had refused corporate funding.

NEJM proclaimed the pro-amoxicillin study the "authorized" version, and the author of the critical study underwent years of discipline and demotions at his university. Five years later, the *Journal of the American Medical Association* published his critical study, and large-scale testing of children showed that those who took amoxicillin actually experienced lower recovery rates than those who took no medicine at all.[8]

In 1989, *The New England Journal of Medicine* came under fire again when it published an article downplaying the dangers of exposure to asbestos but did not inform its readers that the author had ties to the asbestos industry. In 1996, a similar controversy emerged when the journal ran an editorial touting the benefits of diet drugs, again failing to note that the editorial's authors were paid consultants for companies that sell the drugs.

And in November 1997, the *NEJM* published a scathing review of Sandra Steingraber's book *Living Downstream: An Ecologist Looks at Cancer and the Environment.* Controversy erupted after two investigative journalists discovered that the author of the review, Jerry H. Berke, was director of toxicology for W. R. Grace, one of the world's largest chemical manufacturers and a notorious polluter. W. R. Grace is a leading manufacturer of asbestos-containing building products, has been a defendant in several thousand asbestos-related lawsuits, and has paid millions of dollars in related court judgments. It is probably best known as the company that polluted the drinking water of the town of Woburn, Massachusetts, and later paid an $8 million out-of-court settlement to the families of seven Woburn children and one adult who contracted leukemia after drinking contaminated water. During the Woburn investigation, Grace was caught in two felony lies to the U.S. Environmental Protection Agency.

Berke's affiliation with Grace was not disclosed in the *NEJM* review, which ridicules Steingraber as "obsessed . . . with environmental pollution as the cause of cancer" and accuses her of "oversights and simplifications . . . biased work. . . . The focus on environmental pollution and agricultural chemicals to explain human cancer has simply not been fruitful nor given rise to useful preventive strategies. . . . *Living Downstream* frightens, at times misinforms, and then scorns genuine efforts at cancer prevention through lifestyle change. The objective of *Living Downstream* appears ultimately to be controversy."[9]

When questioned about its failure to identify Berke's affiliation with W. R. Grace, the *NEJM* offered contradictory and implausible explanations, attributing the omission at first to an "administrative oversight," claiming that it did not know about Berke's affiliation, and then saying that it did know but thought W. R. Grace was a "hospital or research institute." If so, this ignorance would itself be remarkable, since the *NEJM* is located in Boston and Grace had been the subject of more than a hundred news stories in the *Boston Globe* between 1994 and 1997.[10]

When corporations are not touting the virtues of their products, they use science-for-hire to create seeds of doubt about their hazards. The tobacco industry's extensive history in this regard has been frequently noted, but numerous other examples abound:

• The fossil fuel and automobile industries have spent tens of millions of dollars to create scientific doubt about the real and imminent danger of global warming.

• The food industry has spent tens, if not hundreds of millions, to minimize and deny health risks associated with their products and with the new risks created by factory farming methods.

• With evidence mounting that dioxin and other chlorinated chemicals are carcinogenic, mutagenic and potential endocrine disruptors, the chemical industry has successfully lobbied U.S. government agencies to downplay these hazards, with the result that few Americans have even heard of them.

• Industry-funded science has been used to argue that nuclear power is safe and harmless; that sewage sludge is a valuable fertilizer for human food crops; that genetically engineered hormones in foods are safe and healthy; and that there is no risk to humans from eating foods laced with artificial pesticides and additives such as Monsanto's Aspartame and Proctor & Gamble's Olestra.

"Grassroots" Front Groups

The public relations industry also orchestrates the biggest and most effective so-called "grassroots citizen campaigns" that lobby federal, state, and local governments. Unlike genuine grassroots movements that arise from concerned citizens, corporations control industry-generated "astroturf" movements and pay their bills. PR consultants Edward Grefe and Martin Linsky, in their 1995 book, *The New Corporate Activism,* explain that "The essence of this new way . . . is to marry 1990s communication and information technology with 1960s grassroots organizing techniques." In the 1960s, grassroots activists developed innovative organizing techniques such as the direct-action strategies that Saul Alinsky outlined in *Rules for Radicals*. Today, Grefe and Linsky, whose clients include the tobacco and oil industries, argue that "the heirs of Saul Alinsky can be on both sides of the equation."[11]

The term "astroturf lobbying" describes the synthetic grassroots movements that can be manufactured for a fee by scores of PR firms like Hill & Knowlton, Direct Impact, Optima Direct, National Grassroots & Communications, Beckel Cowan, Burson-Marsteller, Davies Communications or Bonner & Associates. Journalist William Greider has coined his own term to describe corporate grassroots organizing. He calls it "democracy for hire."[12]

"Astroturf" organizing is corporate grassroots at its most deceitful. PR professionals use the term "real grass roots" to refer to orchestrated mass campaigns that are so well designed that they look real. Examples include:

• The Council for Solid Waste Solutions, sponsored by the plastic industry in defense of throwaway plastic

• the Alliance for Responsible CFC Policy, financed by companies such as Dow Chemical to oppose regulation of ozone-destroying chlorofluorocarbons

• the Global Climate Coalition, which lobbies on behalf of the oil, auto, and coal industries to prevent any regulatory interference with global warming

• Citizens for Sensible Control of Acid Rain, which operated between 1983 and 1991 to oppose tightening the Clean Air Act

• the Coalition for Vehicle Choice, created by the Motor Vehicle Manufacturers of America to fight against higher fuel efficiency standards

Organizing from the Bottom Up

Politicians rely on people with money to fund their campaigns. Ultimately, however, they also need votes from the community at large to win election and reelection. Lobbyists therefore need to convince politicians that "the masses" are desperately concerned about the issue they want pressed. By the 1980s, PR firms like Hill & Knowlton were developing techniques not only for targeting legislators but also for serving up their constituents. Since then, the business of organizing grassroots support for pro-business positions has become a half-billion-dollar-a-year PR subspecialty.

The growing proliferation of phony grassroots groups prompted a May 1994 article titled "Public Interest Pretenders" in *Consumer Reports*. "That group with the do-good name may not be what it seems," warned the magazine. "There was a time when one usually could tell what an advocacy group stood for and who stood behind it simply by its name. Today, 'councils,' 'coalitions,' 'alliances,' and groups with 'citizens' and 'consumers' in their names could as likely be fronts for corporations and trade associations as representatives of 'citizens' or 'consumers.' These public interest pretenders work in so many ways—through advertisements, press releases, public testimony, bogus surveys, questionable public-opinion polls, and general disinformation—that it's hard to figure out who's who or what the group's real agenda might be."[13]

As an example, *Consumer Reports* pointed to the Workplace Health & Safety Council, which is actually "a lobbying group composed of employers, and it has opposed a number of regulations aimed at strengthening worker safeguards. Similarly, someone looking at the logo of the National Wetlands Coalition, which features a duck flying over a marsh, would have no clue that the coalition is made up mainly of oil drillers, developers, and natural gas companies."[14]

Stick It Up Your Back Yard

Grassroots organizing is industry's weapon of choice against "NIMBY" or "not in my back yard" movements—local community groups that organize to stop their neighborhood from hosting a toxic waste dump or other unwanted invaders.

John Davies helps neutralize these groups on behalf of corporate clients who include Mobil Oil, Exxon, American Express, and Pacific Gas & Electric. He runs a full-color advertisement designed to strike terror into the heart of even the bravest CEO. It's a photo of the enemy, literally a "little old white-haired lady," holding a hand-lettered sign that reads, "Not in My Backyard!" A caption imprinted over the photo says, "Don't leave your future in her hands. Traditional lobbying is no longer enough. . . . To outnumber your opponents, call Davies Communications."[15]

Davies's promotional material claims that "he can make a strategically planned program look like a spontaneous explosion of community support. Davies has turned grassroots communications into an art form."[16] Davies manufactures friends for needy corporate clients by using mailing lists and computer databases to identify potential supporters.

Pamela Whitney, the CEO of National Grassroots & Communications, also specializes in fighting local community groups. "My company basically works for major corporations and we do new market entries. . . . Wal-Mart is one of our clients. We take on the NIMBYs and environmentalists." National Grassroots also assists "companies who want to do a better job of communicating to their employees because they want to remain union-free."[17]

National Grassroots specializes in "passing and defeating legislation at both the federal and state level," setting up its own local organizations, using a network of professional grassroots organizers. "We believe very strongly in having what we term 'ambassadors' on the ground. One of the things we don't like to do is hire a local PR firm. . . . They are not part of the community. We hire local ambassadors who know the community inside and out to be our advocates, and then we work with them. They report to us."

Flacks to Greens: "Grow Up and Take the Cash"

In addition to creating their own front groups, PR firms also use the strategy of co-optation to recruit potential adversaries into becoming third-party mouthpieces for corporate interests. The February 1994 issue of *O'Dwyer's PR Services Report* gives a candid description of the PR industry's strategy for encouraging sectors of the environmental movement to enter into "partnerships" with major polluters: "The lessons of the recent recession have taught PR people that no matter how idealistic a company sounds, it puts the bottom line ahead of cleaning up its mess," admits an editorial accompanying the report.[18] As a cost-effective alternative, "such companies are finding that cold cash will buy them good will from the environmental movement. Cash-rich companies, PR people say, are funding hard-up environmental groups in the belief [that] the imprimatur of activists will go a long way in improving their reputation among environmentally aware consumers."[19]

On the other side of the "partnership," *O'Dwyer's* observes, "non-profit groups are beginning to realize that private sector cash can increase an organization's clout and bankroll membership building programs." *O'Dwyer's* sees this increased willingness to take "private sector cash" as evidence of "maturing" in the environmental movement.[20]

Dale Didion of Hill & Knowlton in Washington, D.C., one of the nation's largest "environmental PR firms," explains that companies are learning that they can "hire members of the environmental group's staff to help on certain projects. This is a tremendous benefit for a company that wants to have access to top green experts. Companies can avail themselves of talented researchers, scientists and analysts at very reasonable prices."[21]

Starting a relationship between a company and an environmental group carries some risks as well. "It might be in both parties' interest at first to keep their relationship out of the news," says Didion. "Work out early how and when the relationship will be announced to the media and what measure should be taken if word leaks out prematurely," he advises. Didion also suggests some "cost-free and virtually risk free" ways to "test the waters" when entering into a relationship with an environmental group: "Help them raise money. Offer to sit on their board of directors."[22]

Strange Bedfellows

To help industry determine how and which activists can be co-opted, the Public Affairs Council, a trade association for public relations executives, sponsors a tax-exempt organization called the Foundation for Public Affairs. Funding comes from a who's-who list of America's corporate establishment, including Ameritech, Ashland Oil, Boeing, Dow Chemical, Exxon, Health Insurance Association of America, Philip Morris, Mobil, Pharmaceutical Manufacturers Association, RJR Nabisco, and Shell Oil. Many PR-lobby firms also are members, including Bonner & Associates;

Burson-Marsteller; E. Bruce Harrison; The Jefferson Group; and Mongoven, Biscoe & Duchin.[23]

The Foundation for Public Affairs monitors more than 75 specialized activist publications, and gathers information on "more than 1,300 activist organizations, research institutions, and other groups."[24] It publishes an impressive biannual directory titled *Public Interest Profiles,* which offers "intelligence on 250 of the nation's key public interest groups," including "current concerns, budget, funding sources, board of directors, publications, conferences, and methods of operation."[25]

The Foundation for Public Affairs occasionally organizes a Conference on Activist Groups and Public Policymaking, where professional activists and staff members of prominent Washington-based consumer and environmental organizations are invited to rub shoulders with influential corporate PR executives. The conference helps corporate flacks learn how to dissect the strategies, tactics, and agendas of these activists in order to better defeat or co-opt their activism. It is billed as a strictly off-the-record affair, "a one of a kind opportunity to explore the agendas, strategies and influence of leading public interest groups."[26]

Green Backers

In the early 1960s, PR executive E. Bruce Harrison built his reputation by organizing pesticide makers in a PR campaign against environmentalist Rachel Carson and her book, *Silent Spring.* By 1993, Harrison had become a leading "environmental PR" consultant and author of his own book, titled *Going Green: How to Communicate Your Company's Environmental Commitment.* In it, he declares that environmental activism is dead and that its death presents an opportunity to redefine environmentalism in pro-business ways. How did it die? The movement "succumbed to success over a period roughly covering the last 15 years."[27]

After the first Earth Day in 1970, Harrison argues, ecological activism transformed itself from a popular grassroots movement into competing, professionally run nonprofit enterprises—a multimillion-dollar environmental bureaucracy, maintaining expensive offices in downtown Washington and divorced from its activist roots and any meaningful grassroots accountability. With executive directors commanding six-figure salaries, the mainstream environmental organizations are tightly run by boards that increasingly include representatives from *Fortune* 500 companies, including PR firms. These green groups have turned their back on their local supporters, who are little more than recipients of cleverly worded junk mail funding appeals.

Going Green says that despite their formal nonprofit status, today's big environmental groups are first and foremost business ventures. Harrison advises his PR clients that the green groups' real goal "is not to green, but to ensure the wherewithal that enable it to green." The managers of the big green groups are primarily concerned with raising money from individuals, foundations, and increasingly from corporations. To do so they have chosen to maintain a "respectable" public image and are very willing to work with industry PR executives. This puts mainstream environmental groups in a position to be compromised through industry partnerships and funding.

Some of the biggest and best-known green organizations—the Wilderness Society, the National Wildlife Federation, and the National Audubon Society among them—are receiving support, recognition, and large cash contributions from corporate polluters. In exchange, the corporate benefactors have been able to buy a green image that is worth literally millions in the consumer marketplace. Harrison's PR firm spends much of its time helping its Fortune 500 clientele build issue coalitions, partnerships, and alliances with carefully chosen pro-business environmentalists.

As an example of an ideal partnership, Harrison points to the marriage between McDonald's restaurants and the Environmental Defense Fund (EDF). After the Citizens Clearinghouse on Hazardous Waste organized a national grassroots campaign against McDonald's use of plastic foam containers, EDF Executive Director Fred Krupp negotiated a highly publicized settlement and began an ongoing "partnership" with the fast-food behemoth. Krupp gained a "victory," which the EDF highlights prominently in its fund-raising activities. Bragging about this achievement has helped EDF raise millions of dollars.

Harrison provides a bottom-line assessment of the deal: "In the late 1980s, [McDonald's] slipped into its worst sales slump ever—and the anti-McDonald's drive of the [grassroots] green activists was at least partly blamed. . . . Krupp saw the golden arches of McDonald's, the nation's fast food marketing king, as a sign of opportunity. . . ."[28] EDF's mission, Krupp said, is not to attack corporations but "to get environmental results." He told the *New York Times,* "Being willing to consider new ways to regulate and being willing to talk with business in a business-like way is not the same as being in favor of halfway compromises."[29]

The main beneficiary of the agreement, however, has been McDonald's, which saw its environmental reputation soar. Opinion polls following the deal gave McDonald's one of the highest environmental ratings of any U.S. corporation.

Meanwhile, McDonald's remains a massive corporate polluter. It continues to hire and underpay an unorganized work force that sells greasy, fatty food grown with pesticides on factory farms, sold by an international franchise that destroys community economic diversity, and that advertises with billions of dollars that target children using bigger-than-life millionaire sports celebrities. When a tiny group of London environmental activists had the audacity to leaflet McDonald's and publicly criticize its destructive policies, the fast food giant sued them under Britain's reactionary libel laws to prevent even this small mention of the truth from undermining the company's cultivated image.[30]

In *Losing Ground,* author Mark Dowie cites the EDF–McDonald's arrangement as an example of "high-level capitulations" that unfortunately "allow companies like McDonald's to look a lot greener than they are. The corporate exploitation of so-called 'win/win' compromises has been relentless. Companies compete through paid and free media to out-green one another. . . . It's predictable and understandable. But the environmentalists' complicity and their own PR-driven tendency to turn compromise into false triumph illuminates the desperation and impending moral crisis of the mainstream organizations."[31]

At the grassroots level, meanwhile, thousands of citizens are engaged in genuine environmental activism, facing off in their communities against the waste dumpers, wanton developers, and pesticide pushers. These grassroots activists are being outmaneuvered by the big green groups, which soak up almost all the environmental money from green philanthropists and small donors alike, while providing little or no support to the legitimate foot soldiers of the environmental movement.

The Good Guys Guise

"There has recently been a spurt of corporate advertising about how corporations work to clean the environment," writes Jerry Mander, author of *In the Absence of the Sacred*. "[C]orporations will tend to advertise the very qualities they do not have, in order to allay a negative public perception. When corporations say 'we care,' it is almost always in response to the widespread perception that they do not care."[32]

Some of the industrial polluters with the worst records have devised "public education" campaigns that enable them to placate the public while they continue polluting. The agrichemical conglomerate Monsanto, for example, has given away hundreds of gallons of its RoundUp herbicide through "spontaneous weed attack teams" (SWAT) for spraying in inner-city neighborhoods to make them "cleaner and safer places to live."

Dow Chemical's environmental PR campaign began in 1984 with the goal of making "Dow a more highly regarded company among the people who can influence its future." Dow's reputation was still suffering from its manufacture of napalm bombs and the Agent Orange defoliants that devastated much of Vietnam. The company mailed glossy "public interest reports" to 60,000 opinion-makers: scientists, the media, legislators, regulators, employers, customers, and academics. Illustrated with numerous high-quality photographs, the "public interest reports" touted Dow's programs in the area of environment and five other "good works" categories.[33]

In 1986 a poll by the *Washington Journalism Review* found that business editors rated Dow's PR efforts tops among *Fortune* 500 chemical companies. As a member of the Chemical Manufacturers Association, Dow participates in Responsible Care, a PR program in which each chemical company evaluates its own environmental performance. Its advertising slogan reinforces the message: "Dow helps you do great things." As a result of this systematic campaign, *American Demographics* listed Dow in 1993 as one of the ten U.S. firms with the best environmental reputations among consumers.[34]

"Many people use [Dow] as an example of doing the right thing. There is hardly a discussion of pollution control and prevention among American industries that fails to highlight Dow and the strides it has made," writes Jenni Laidman in the *Bay City Times* of Saginaw, Michigan. Laidman notes that Dow garners all this praise even though the company "is still a leading polluter in the state and the nation. . . ."[35]

In addition to co-opting environmental moderates, corporate PR firms are helping companies set up "community advisory panels" (CAPs) to strengthen their image in neighborhoods that contain industrial facilities. Dow Chemical is one of the companies that has pioneered the establishment of CAPs. "[CAPs] will be an integral part of doing business in all major industries," said A. J. Grant, president of Environmental Communication Associates in Boulder, Colorado. "You've got to have a marketing department, you've got to have accounting, and you'll have to have community interaction in the form of a CAP."[36] According to Joel Makower, editor of *The Green Business Letter*, CAPs "differ in makeup, style, and function," but "a typical CAP includes activists, homemakers, community leaders—a representative sampling of just plain folks—as well as company representatives." CAPs create a forum for dialogue between the company and the community, but the company carefully modulates the nature of the dialogue to emphasize emotions and image shaping rather than issues of substance.

As an example of their PR effectiveness, Makower relates the following anecdote: "Members of one CAP, unbeknownst to the company, appeared voluntarily before a local hearing to testify why the company should be allowed to site an incinerator in their backyard. You can't buy that kind of help at any price."[37]

Is Resistance Futile?

Any analysis of the power and scale of the modern public relations industry seems overwhelming and disempowering upon first examination. However, this power should be viewed in its historical context. Alex Carey observes that "the twentieth century has been characterized by three developments of great political importance: the growth of democracy, the growth of corporate power, and the growth of corporate propaganda as a means of protecting corporate power against democracy."[38]

Indeed, the growth of PR has been largely a reactive phenomenon to corporate fears raised by popular democratic movements. This fact has been frequently noted within the industry itself ever since the public relations profession began as a response to challenges that the Progressive and labor movements posed to corporate power. Fear of popular movements remains a pressing concern for PR professionals even today, notwithstanding the seeming global ascendancy of corporate power. "The public has turned against corporate America now more than at any time since the 1960s," observed leading Republican Party consultant Frank Luntz in a confidential 1996 memo. "This time the frustration and anger stretches well into the middle class and up through white collar and middle level management," Luntz added.[39]

As every corporate lobbyist and public affairs officer understands, the public is increasingly unhappy with the consequences of corporate "bottom-line" behavior—consequences such as layoffs, forced production speedups, union busting, deregulation, wage and benefits cuts, reductions in government services, unravelling consumer and environmental safeguards, economic and racial polarization, global sweatshops, monopolization, and price fixing. Opinion polls in 1991 showed that 69 percent are concerned about "worsening social problems resulting from growing numbers of poor people," and many believe "the free market system is not fair." Seventy-nine percent believe that the government is "run by a few big interests looking out for themselves." Fifty-nine percent said there was not a single elected official today that they admired. Forty-six percent believe that the middle class is being hurt by "corporate greed."[40]

The PR profession, founded to contain and control public opinion, faces major image problems itself and is almost universally perceived as a major source of sleaze, "spin," and insincerity. "Front groups are beginning to wear out their welcome. Increasingly they are being 'outed' by legitimate activists," complained the summer 1996 newsletter of the Public Affairs Council.[41]

The three main threats mentioned most frequently by PR professionals are trade unions ("big labor"), trial lawyers (who sue corporations), and the news media. Corporations fear the news media because they believe that it has become so ratings conscious and sensationalistic that it is literally "out of control." Facing intense competition for audience share, media organizations sometimes run stories with popular appeal, even when they present big business interests in an unfavorable light. Public relations practitioners worry that this media environment, combined with today's cynical public mood, creates opportunities for its enemies—unions, trial lawyers, the environmentalists, and "NIMBY" groups. Corporate PR operatives even fear that their own innovative use of ever-more-affordable information technology—e-mail, patch-through 800 lines, sophisticated polling, broadcast faxes, internet Web sites—could become a double-edged sword as activists learn to use the same techniques against them.

Public relations practitioners frequently mention in their own defense that many PR strategies are simply extensions of democratic practices—lobbying, communications, grassroots organizing. It would be impossible and wrong to legally ban such activities, because any laws that attempt to do so would inevitably inflict severe harm on the rights of genuine citizen activists.

Grassroots activism, even when practiced by a handful of citizens, is already a limiting factor on the success of corporate propaganda. That corporations feel compelled to seek "partnerships" with environmentalists is evidence of their realization that they cannot rule by fiat. Indeed, many public relations professionals admit frankly that their main purpose in pursuing such partnerships is to dissuade activists from pursuing the aggressive tactics of grassroots activists, which they perceive as environmentalists' strongest weapons. Harrison writes that "Greening and the public-policy impact of greenism are being propelled by what I refer to as the 'AMP Syndrome' a synergy of Activists + Media + Politicians. . . . Activists stir up conflict, naming 'victims' (various people or public sectors) and 'villains' (very often, business interests). The news media respond to conflict and publicize it. Politicians respond to media and issues, moving to protect 'victims' and punish 'villains' with legislative and regulatory actions."[42]

The industry strategy of organizing grassroots movements in order to counteract the work of other grassroots movements may ultimately prove self-defeating. History offers numerous examples of social movements which, once undertaken, develop autonomy and goals quite different from the goals that marked their founding. For this reason corporate grassroots strategist Neal Cohen warns PR professionals to maintain a distinction between "broad-based membership" and tightly centralized decision-making: "Broad-based membership is: What does the public see? What do the legislators see? Decision-making is: a core group of three or so people who have similar interests and who are going to get the job done and not veer off."[43]

Whether and how long the PR industry can remain successful in keeping its publics from "veering off" in uncontrolled directions remains unclear. However, the industry's fear of such contingencies suggests that radical, idealistic grassroots organizing by uncompromising activists offers the best hope we have for a genuinely democratic future.

Notes

1. Alex Carey, *Taking the Risk Out of Democracy: Propaganda in the US and Australia* (Sydney, Australia: Univ. of New South Wales Press, 1995, p. 12).

2. Susan B. Trento, *The Power House: Robert Keith Gray and the Selling of Access and Influence in Washington* (New York: St. Martin's Press, 1992, p. 62).

3. Edward L. Bernays, *Propaganda* (New York: Horace Liveright, 1928, pp. 47–48).

4. Merrill Rose, "Activism in the 90s: Changing Roles for Public Relations," *Public Relations Quarterly* 36(3) (1991):28–32.

5. Joel Bleifuss, "The Truth Hurts," *PR Watch* 3(4):1, 1996.

6. Cynthia Crossen, *Tainted Truth: The Manipulation of Fact in America* (New York: Simon & Schuster, 1994, pp. 183–84).

7. Ibid., p. 42.

8. Ibid., pp. 161–65.

9. Jerry H. Berke, "Living Downstream" (book review), *New England Journal of Medicine* 337 (1997):1562.

10. Nate Blakeslee, "Carcinogenic Cornucopia," *Texas Observer,* January 30, 1998.

11. Edward A. Grefe and Martin Linsky, *The New Corporate Activism* (New York: McGraw-Hill, 1995, pp. xi, xiii, 2).

12. William Greider, *Who Will Tell The People* (New York: Simon & Schuster, 1992, p. 35).

13. "Public Interest Pretenders," *Consumer Reports* (May 1994):316.

14. Ibid.

15. Advertisement, *Campaigns & Elections* (December/January 1995):4.

16. John Davies speaking at "Shaping Public Opinion: If You Don't Do It, Somebody Else Will," in Chicago (December 9, 1994).

17. Pamela Whitney at "Shaping Public Opinion: If You Don't Do It, Somebody Else Will," in Chicago (December 9, 1994).

18. "Green PR Is Dollars and Sense Issue," *O'Dwyer's PR Services Report* (February 1994):6.

19. "Links with Activist Groups Get Results in Environmental PR," *O'Dwyer's PR Services Report* (February 1994):1.

20. Ibid.

21. Ibid., p. 20.

22. Ibid., p. 22.

23. "Countdown to Our Fifth Decade," The Public Affairs Council (1993): 18–23.

24. Foundation for Public Affairs, 1993–1994 annual report, p. 2.

25. Foundation for Public Affairs, 1993–1994 annual report, p. 5.

26. Conference call, Annual Conference on Activist Groups and Public Policymaking: Agendas, Strategies, and Alliances with Business, Washington, D.C. (October 20–21, 1993).

27. E. Bruce Harrison, "Managing for Better Green Reputations," *International PR Review* **17**(3) (1994):8.

28. Ibid., p. 216.

29. Keith Schneider, "For the Environment, Compassion Fatigue," *New York Times*, November 6, 1994.

30. Tom Kuntz, "The McLibel Trial," *New York Times*, August 6, 1995, p. E7.

31. Mark Dowie, *Losing Ground: American Environmentalism at the Close of the 20th Century* (Cambridge, Mass.: MIT Press, 1995, p. 140).

32. Jerry Mander, *In the Absence of the Sacred: The Failure of Technology and the Survival of the Indian Nations* (San Francisco: Sierra Club Books, 1991, p. 131).

33. Allen Center and Patrick Jackson, *Public Relations Practices*, 4th ed. (Englewood Cliffs, N.J.: Prentice-Hall, 1990, p. 354).

34. Peter Stisser, "A Deeper Shade of Green," *American Demographics* (March 1994:28.

35. Jenni Laidman, *Bay City Times*, Saginaw, Mich. (September 12, 1994).

36. *The Green Business Letter*, 1(March 1994):6–7.

37. Ibid.

38. Carey, *Taking the Risk Out of Democracy*.

39. "They're Rich, They're Powerful, and They're Running Scared," *PR Watch* **4**(1) (1991):1

40. Ibid.

41. *Impact* (newsletter of the Public Affairs Council) (Summer 1996).

42. Harrison, "Managing Green Reputations," p. 277.

43. Neal Cohen speaking on "Coalitions and Ally Development: The New Imperative in Public Policy Work" at Public Affairs Council conference, Sarasota, Fla. (February 7, 1994).

10

The Globalization of Corporate Culture and Its Role in the Environmental Crisis

Joshua Karliner

While the 1990s were supposed to be the environmental decade, few people realized that the most visible and aggressive ecological protagonists at the end of the twentieth century would be the global corporations. A generation ago, when astronauts first beamed pictures of the Earth back home, the image of a life-sustaining blue planet floating in infinite space helped inspire social movements to work to save the world from the depredations of industrialism. Yet today the Earth's image has been co-opted into a corporate logo and is found in advertisements for enterprises ranging from communications companies to fast-food dealers to chemical giants. Transnational corporations elevate the blue planet as an object whose transcendent problems of ecological decay and pervasive poverty can be solved only by the technological, economic, and managerial prowess at their command—a capacity unmatched in human history.[1]

The stunning portrait of the Earth has also become the icon for a millennial version of manifest destiny known as globalization. We are told that this corporate encirclement of the planet will bring with it greater prosperity, peace, and ecological balance. It may be difficult to resist embracing such an enticing vision, especially without a clear counterbalance or alternative on the horizon. Soviet and Eastern European communism has left little but a legacy of ruin (including ecological devastation). In many respects traditional nation-states, including both the high-tech industrial democracies and the multitude of third world governments, are growing weaker and less relevant. At the same time, a growing number of countries are experiencing escalating nationalism and ethnic division, leading to xenophobia, fundamentalism, fascist tendencies, and war. This leaves the corporate capitalists and the leaders of the industrialized democracies free to present their brand of globalization as a historical inevitability and themselves as healers of the world's ills.

In the absence of a coherent alternative, the transnational corporations, which are increasingly flagless and stateless, weave global webs of production, commerce, culture, and finance virtually unopposed.[2] They expand, invest, and grow, concentrating ever more wealth in a limited number of hands. They work in coalition to influence local, national, and international institutions and laws. Together with the governments of their home countries in Europe, North America, and Japan, as well as international institutions such as the World Trade Organization, the World Bank, the International Monetary Fund, and increasingly, the United Nations, they are molding an international capitalist system in which they can trade and invest even more freely. Underpinning this effort is not the historical inevitability of an evolving, enlightened civilization, but rather the unavoidable reality of the overriding corporate purpose: the maximization of profits. The corporate planet is encroaching upon the blue planet, commodifying it, homogenizing it, and enclosing it within its predatory global reach.

The "globalization" that has taken place in the 1990s is an acceleration of historical dynamics, hastened by the advent of increasingly sophisticated and rapid communications and transportation technologies, the decline of the nation-state, the absence or ineffectiveness of democratic systems of global governance, and the rise of neoliberal economic ideology. Indeed, it is highly unlikely that *corporate globalization* (or, perhaps better said, global corporatization) will foster a world based on democratic participation, social justice, nonviolence, healthy communities, and ecological sustainability.

Given their penetrating reach, it is not surprising to find the big transnationals deeply involved in most of the world's serious environmental crises. These corporations effectively play the role of Earth brokers in the global economy—buying and selling the planet's resources and goods, deciding what technologies will be developed and used, where factories will be built, which forests will be cut, minerals extracted, crops harvested, and rivers dammed. Their inordinate power puts them in the position of determining the future of local, regional, and global ecosystems in the interests of an antiquated version of economic growth that is wreaking havoc on the world's ecology and public health.

The 1992 Earth Summit in Rio de Janeiro marked the coming of age of corporate environmentalism—the melding of ecological and economic globalization into a coherent ideology that has permitted the transnationals to reconcile, in theory and rhetoric, their ubiquitous hunger for profits and growth with the stark realities of poverty and environmental destruction. In the aftermath of Rio, global corporate environmentalism has helped institutionalize ecological concerns as agenda items and built a public image of transnational corporations as the world's responsible global "citizens." It has also, to a certain degree, set the terms of the debate along lines favorable to the transnationals. In the process, corporate environmentalism has partially neutralized efforts—ranging from popular environmental movements to intergovernmental treaties and conventions—that pose a threat to their power.

They have appropriated the language and images of ecology and sustainability in an effort to ward off the threat that the environmental movement might convince the world's governments to force them to make much more far reaching changes. Self-proclaimed corporate environmentalists achieved this by absorbing ecological sustainability into their overriding agenda of economic globalization. Suddenly they have made the worldwide expansion of resource extraction, production, marketing, and consumption synonomous with sustainable development.

The ideological underpinnings of corporate environmentalism were most clearly articulated by the Business Council for Sustainable Development at the 1992 Earth Summit, when they released the book *Changing Course*. A reading of this tome reveals that corporate environmentalism rests on four pillars. First it contends that unleashing market forces to promote continuing economic growth through open and competitive trade is the fundamental prerequisite for sustainable development. Second, it calls for pricing mechanisms to correct distortions in the world economy and reflect environmental costs. Third, it argues that self-regulation is the best, most preferable, and most efficient method for transforming business practices. And fourth, it calls for more changes in technology and managerial practices in order to promote cleaner production and the more efficient use of resources.[3] These pillars, erected in

1992, continue to serve as the fundamental framework for corporate environmentalism today.

This approach to solving the world's social and environmental problems envisions a rosy scenario of more growth through globalization, more profits, and a healthy corporate planet. Yet it is problematic; its technocratic, top-down approach tends to undermine democracy and community environmental rights.

Under any circumstances, corporations could certainly assume far greater responsibility for their actions than they have. It is important to recognize that *some* change is beginning to take place. There are many socially and ecologically conscientious individuals who work for these companies. Many of them are working to alter the corporations that employ them on every level, from the shop floor to the boardroom. These innovations and reforms have manifested themselves in a number of concrete ways. For instance, while environment was a topic rarely taken seriously in the business world, since 1985 a plethora of corporate environmental departments and policies have emerged to set internal standards, rules, guidelines, and long-term goals. Japanese corporations, for instance, set up more than 300 environmental affairs departments in the early 1990s alone. It is also not uncommon today to encounter people in charge of environmental issues among the ranks of senior executives and members of the board of directors of a transnational. Moreover, global environmental management has become a respected profession, while environmental self-audits and environmental impact assessments have become a common tool for many transnational corporations.[4]

Industrywide voluntary codes of conduct have also proliferated in the past decade or so. A number of corporations are beginning to explore how to account for environmental costs in their economic calculations.[5] A plethora of positive case studies can be found touting changes in corporate management, as well as technological innovations that are reducing waste, minimizing the impact of destructive production processes, and making companies more environmentally sound in general.[6]

These various initiatives have evolved, however, within the context of economic globalization, which is promoting accelerated environmental destruction along with social dislocation. Global corporations are molding a world in which they are less accountable to the law of nations, cultures, or communities. In place of these traditional institutions, transnationals are creating methods of self-governance and autoaccountability designed to guide their interactions with the people they affect.

In this new pseudosystem of corporate governance, such people are envisioned as "stakeholders" in a corporate enterprise. Stakeholders range from first-class shareholders, customers, and employees to second- or third-class advocacy groups and communities who might live, for example, next to a polluting factory. The designation of "stakeholder" tends to remove citizens from the realm of political power achieved through participation in democratic institutions (if their society has such institutions). It also overrides any traditional community decision-making process and resource-use patterns. In place of these conventions, it redefines citizens and their communities as constituencies of transnational corporations in the world economy, virtually defining them as residents in a global version of the "company town."

On the global level, corporate environmentalism strongly endorses international economic agreements such as the North American Free Trade Agreement (NAFTA), the World Trade Organization (WTO), and the proposed Multilateral Agreement on Investment (MAI) as essential to addressing ecological questions. Indeed, it is virtually a mantra of corporate environmentalism that "open trade is a key requirement for sustainable development."[7] However, as former Harvard Business School professor David Korten argues, by playing an integral role in drafting and supporting the passage of these agreements, the transnationals are "actively rewriting the rules of the market to assure their own rights and freedoms take precedence over the rights and freedoms of the world's citizens."[8] The environmental implications of these agreements are serious. Obscure bureaucratic decision-making panels are empowered to overturn local and national laws protecting the environment and consumers' and workers' rights. For instance, the WTO has ruled that a Canadian fisheries conservation measure, a Thai law limiting cigarette imports, and U.S. laws that taxed oil and chemical feedstocks to pay for hazardous waste cleanup are unfair barriers to trade.[9]

These agreements are also having a chilling effect on the formulation of future environmental and other policies by limiting governments to approaches that are consistent with WTO rules.[10] Meanwhile, mechanisms in international environmental agreements such as the Montreal Protocol to Protect the Ozone Layer, which provide for the transfer of ozone-friendly technologies to the third world, are also potentially in conflict with the WTO, which may view such measures as unfair subsidies.[11]

Perhaps the most striking indictment of corporate environmentalism can be found in comparing the transnationals' words with their actions. These contradictions were most apparent when the Republican party swept into control of the 104th U.S. Congress in 1995. Suddenly, "environmentalist" companies such as Chevron, Dow Chemical, and the timber transnational Georgia Pacific, all of whom sat on the Clinton administration's President's Council on Sustainable Development (PCSD),[12] showed their true colors by participating in an all-out assault to paralyze and gut U.S. environmental regulations.[13]

For instance, together with their Republican allies in the House and Senate, U.S. corporations opened up ancient forests on federal lands to clear-cutting, passing a law (signed by President Clinton) that overrode environmental protections already in place.[14] While many of their initiatives were stymied in the legislative process, or by presidential veto, the scope of the attempted rollback effort was truly stunning. They worked to slash the budgets for enforcement agencies such as the Environmental Protection Agency and the Interior Department.[15] As they continued to deny the scientific basis of many environmental problems, they set their sights on making significant cuts in nearly all federal environmental science programs that provide crucial data on problems such as global warming.[16] They attempted to undermine laws mandating clean air and water, as well as those protecting endangered species and wilderness.[17] They moved to delay the phaseout of ozone-depleting chemicals.[18] They sabotaged food safety laws.[19] They attempted to undermine laws that make them liable for defective products, as well as those that force them to pay for pollution crimes.[20]

U.S. corporations also helped write and promote laws that use a risk assessment formula to make economic considerations the determining factor over health protection when setting environmental standards, thus undermining the government's ability to regulate corporate activities in the public interest. Risk assessment also promised to tie up government enforcement of environmental regulations in red tape. U.S. corporations offered a takings law that inverts the burden of compensation, requiring that the government reimburse landowners (often corporations) if a law lowers their property values. For instance, the federal government would be required to compensate a land-owning corporation for "taking" its property if the government denied it a permit to build a hazardous waste dump over a shallow aquifer. Meanwhile the "landowner" might not be forced to compensate the community whose aquifer it poisons if a risk assessment determined that the economic value of the operation was greater than the cost in human health.[21]

The Emerald City: Advertising, Public Relations, and the Production of Desire

In many respects, transnational corporations have gone to great lengths to conjure up a mirage of sustainability rather than transform their practices. In an effort to appropriate the symbols, language, and message of environmentalism while continuing to promote ever-expanding consumerist societies, they have built a dream world of image and myth—a global Emerald City.

The corporate world seeks to market itself and its products as the greenest of the green. One-fourth of all new household products that came onto the market in the United States around the time of Earth Day 20 advertised themselves as "recyclable, "biodegradable," "ozone friendly," or "compostable."[22] Simultaneously, some of the world's greatest polluters spent millions putting on a shiny new coat of green paint—both literally and figuratively. The oil company ARCO, for instance, concealed its Los Angeles oil facility behind a facade of palm trees and artificial waterfalls in what one commentator labeled an "industrial version of cosmetic dentistry."[23] Dow Chemical, the largest producer of chlorine in the world, used the image of the planet Earth to tout its "ongoing commitment" to the environment, which it claims can be traced to the founding of the company.[24]

Similarly, across the Pacific, nuclear giant Hitachi was billing itself in advertisements as "a citizen of the Earth." And a Mitsubishi Corporation joint venture that clear-cut vast swaths of hundred-year-old aspen forests in Canada, producing between six and eight million pairs of disposable chopsticks a day, exported them to Japan, where they were sold as "chopsticks that protect nature."[25]

In Europe, greenwashing was no less prevalent. The Swiss chemical corporation Sandoz, in an effort to rehabilitate its image after the 1986 Basel chemical spill, ran advertisements depicting a forest, a tranquil pond, and a clean river running through the scene. To a certain degree, the advertisement was accurate; by 1990 Sandoz had relocated its hazardous chemical production from Switzerland to Brazil and India.[26]

This toxic greenwash also spilled into the third world. In Malaysia, for instance, the British corporation ICI produced a blatantly deceptive full-color newspaper advertisement whose headline trumpeted "Paraquat and Nature Working in Perfect Harmony."[27] The ad, which described Paraquat as "environmentally friendly," contained a series of outrageous assertions about this highly toxic herbicide, which has poisoned tens of thousands of workers in Malaysia alone, is banned in five countries, and is listed as one of the "dirty dozen" by the Pesticide Action Network.[28]

In 1985 Chevron launched its People Do advertisements. Still going strong many years later, the People Do series is a textbook case of successful greenwashing. It began when Chevron asked itself whether it would pay to tailor an advertising campaign to a "hostile audience" of "societally conscious" people concerned about such issues as offshore oil drilling.[29]

Produced at a cost of $5 to $10 million a year, the campaign consists of an expanding series of advertisements, each of which features a different Chevron People Do project. For instance, in one of the first People Do ads, a tiny blue butterfly flutters across a television screen and into the homes of millions of viewers. A narrator's voice explains Chevron's program to save this endangered species, and rhetorically asks his audience, "Do people really do this so a tenth of a gram of beauty can survive?" The narrator then drives home the theme of the advertising campaign with the slogan: "People Do." However, at the time the ad was produced, the sand dune–dwelling El Segundo blue butterfly made its home within a barbed wire–fenced compound on top of the United States' largest underground oil spill. The viewer is not aware that the butterfly flutters within the confines of one of the largest sources of pollution in the Los Angeles Basin—Chevron's El Segundo refinery.

The television and magazine spots, which are designed to recast Chevron's image, create the impression that Chevron, one of the world's major polluters, is a caring, clean, green, butterfly-loving group of people.

In addition to the butterfly "preserve" at the El Segundo refinery, the People Do ads have publicized artificial reefs made of old gas station storage tanks the company sank off the coast of Florida, its efforts to protect grizzly bears near one of its drilling sites in Montana, and artificial kit fox dens in the Central Valley of California. Critics charge that the ads are misleading. Herbert Chao Gunther, director of the Public Media Center in San Francisco, comments that "the ads are a selective presentation of the facts with a lack of context. Chevron implies that maybe we don't need a regulatory framework because the oil companies are taking care of it."[30]

It appears that one major factor motivating Chevron's People Do campaign is the transnational's deregulatory agenda. An investigation of People Do by a local San Francisco television station discovered that although Chevron sells gasoline across much of the country, the corporation has aired its ads only in the top three oil-producing states of the continental United States—California, Texas, and Louisiana—locations where it drills for and refines most of its oil, and consequently where it is most heavily regulated. Confronted with this evidence, Chevron spokespeople insist that People Do is not a "political advocacy program" (if it were found to be so, their advertising might no longer be tax deductible). However, the only other place in the United States where Chevron airs its People Do advertisements is in Washington, D.C., which is hardly a nationally significant gasoline market.[31]

Despite public skepticism, the People Do strategy seems to have worked. Polls that Chevron conducted in California two years after the campaign began show that it had become the oil corporation people trusted most to protect the environment. Among those who saw the commercials, Chevron sales increased by 10 percent, while among a target audience of the potentially antagonistic, socially concerned types, sales jumped by 22 percent.

The U.S. Federal Trade Commission has issued a series of "green" guidelines aimed at halting deceptive greenwash ads.[32] In 1990, a task force of eleven state attorney generals issued a report that called for environmental claims to be "as specific as possible, not general, vague, incomplete or overly broad."[33] In France and the Netherlands, environmentalists have introduced a twelve-point advertising code to their national legislatures. Independent green labeling regimes have also emerged in a number of countries. The best of these efforts attempt to create independent criteria to evaluate the environmental effects of a product throughout its life cycle. However, despite these initiatives, as media ownership becomes increasingly concentrated in the hands a few giant corporations, and as economic globalization promotes ever-expanding media reach into the growing consumer markets of the South and the East, greenwash, along with the ecologically unsound consumption patterns that accompany it, continues to flow.

Global Images and the Production of Desire

The profound influence that corporations wield over the words and images fed to the general public seriously affects how we live and how we understand the nature of the environmental crisis and the changes needed to correct it. Direct corporate control of the media has become increasingly concentrated since World War II. By 1982, just fifty corporations controlled more than half of all major U.S. media, including newspapers, magazines, radio, television, books, and film. By 1993 this number had shrunk to fewer than twenty.[34] By the mid- and late 1990s, the trend had intensified as U.S. telecommunications, computer, and media corporations merged and acquired each other at a breakneck pace.

Despite protests of objectivity by media leaders, this centralization of ownership results in severely skewed reporting that undermines democratic debate and the role it should play in promoting ecological sustainability. For instance, when NBC ran a documentary praising nuclear power (without mentioning that its parent company, General Electric, is a major nuclear power plant builder), or when the network's news division ran a series of segments about a new device to detect breast cancer (without mentioning that the device is manufactured by GE), the supposedly "objective" media came very close to resembling a glorified corporate public relations and marketing operation.[35]

Television networks and news magazines regularly refuse to run stories on the social or environmental impacts of a particular product, corporation, or industry. Such coverage would jeopardize their portion of the more than $3 billion that corporations spend every year on advertising in U.S. media markets alone.[36]

By promoting the same products, role models, and values across the planet, television advances the global sprawl of a homogeneous culture of consumption, what Vandana Shiva has referred to as a "monoculture of the mind," that is ravaging and replacing more than 6,000 cultures around the world. This ethic of boundless consumption underpins the accelerating depletion and despoliation of the world's natural resources. In the South and the East, people are abandoning traditional, often sustainable ways of life in favor of consuming resources extracted, manufactured, and marketed by transnational corporations. While the northern industrialized countries do little to reduce their profligate consumption of resources, the globalization of American consumption patterns lures more people into participating in an utterly unsustainable life that is already destroying the planet's life support systems. In this way the destruction of the world's cultural diversity is intertwined with the destruction of the world's biological diversity—and with it the ecological equilibrium of the Earth.

The Wizards of PR

The few corporations that dominate the public relations industry are playing an increasingly pivotal role in environmental politics around the world. (See the chapter by Rampton and Stauber for a fuller discussion of the activities.) "PR experts—at Burson-Marsteller, Ketchum, Shandwick, and other firms" write Stauber and Rampton, "are waging and winning a war against environmentalists on behalf of corporate clients in the chemical, energy, food, automobile, forestry and mining industries."[37] Public relations corporations, many of which are subsidiaries of the world's largest advertising agencies, manage what they call "integrated communications" strategies. These schemes regularly combine the use of slick greenwash ads and the placement of "real" news stories in the media, with a series of services that range from "green" marketing plans to espionage and "crisis management" to the orchestration of fake grassroots or "astroturf" campaigns to highbrow lobbying.

While corporations, industry groups, and PR firms work to fend off movements for greater public control and accountability, they are also infiltrating educational systems in an effort to preempt the emergence of a new generation of critical environmental activists. Such initiatives, which date back to the 1970s in the United States, have picked up steam in the 1990s, while also spreading to other countries.

The greatest culprit in this endeavor is Exxon, which is rewriting the history of the Valdez oil spill for an audience of the nation's impressionable youth. While an Alaska jury was awarding thousands of Native Alaskans and another 20,000 plaintiffs more than $5 billion in damages from Exxon, the company was distributing, free of charge, its version of the truth to 10,000 elementary school teachers, for viewing by children who were too young to remember the devastating oil spill. While the jury determined that the spill had destroyed much of the plaintiffs' livelihood—damaging fishing and native hunting grounds, the Exxon video told a new generation of potential environmentalists and soon-to-be-consumers that the spill did not decimate wildlife in Prince William Sound.[38]

Exxon is only one participant in the corporate invasion of public schools. As the *Philadelphia Inquirer* puts it, "The next time your child's class turns to its lesson plan on the environment, it is likely that the materials the teacher passes out will have been supplied by Procter & Gamble . . . or Browning Ferris Industries . . . or Exxon, Chevron or Mobil."[39] For instance, the Procter & Gamble program, Decision Earth, which was distributed to 75,000 schools, touts clearcut deforestation as ecologically beneficial because it helps "create new habitat for wildlife."

The American Coal Foundation provides a curriculum that makes no mention of acid rain or global warming, but rather helps students "identify the reasons coal is a good fuel choice." The American Nuclear Society supplies teachers with a "science/social studies fair kit" that provides step-by-step instructions for building a model nuclear reactor. The society's kit writes off the problem of nuclear waste, telling students that "anything we produce results in some 'leftovers' . . . whether we're making electricity from coal or nuclear, or making scrambled eggs!"[40]

Another corporate PR strategy has been to divide and conquer the U.S. environmental movement. First, it promoted "partnerships" between large polluting corporations and the most establishment-oriented environmental organizations; this had the effect of co-opting or influencing to a degree some of the mainstream groups, while simultaneously driving a wedge between these organizations and the more progressive elements of the environmental movement. At the same time, it pursued a second track aimed at fomenting an antienvironmental backlash.

Declaring an end to the era of confrontational politics, PR firms geared up in the early 1990s to promote cooperation and "harmony" among adversaries in the environmental debate. "The Cold War is over between environmental activists and companies" trumpeted *O'Dwyer's PR Services* in 1994. "Each side now is willing to work with the other on projects designed to improve the environment." Such collaboration, *O'Dwyer's* observed, helps improve a corporation's reputation among environmentally aware consumers, while thickening the cash flow for environmental organizations.[41]

Mainstream environmental leaders, such as World Resources Institute (WRI) President Jonathan Lash, insist that in order to achieve "sustainable development" there is "no choice but to take a collaborative approach." He compares the growing harmony between mainstream environmental groups, corporations, and government with the end of the Cold War and the "velvet revolutions" of Eastern Europe, as well as with the African National Congress' ascent to power in South Africa. Lash suggests that just as change was achieved "without violence" in these situations, the U.S. government, corporations, and the mainstream environmental groups are working together to stave off the threat of "a much more confrontational approach" to addressing the environmental question.[42]

While Lash proposes close cooperation among corporations, environmental organizations, and government, some critics see this approach as a deception and a sellout. Many grassroots organizers insist that violence is already being done by the corporations. Lois Gibbs, director of the Center for Health, Environment and Justice, whose activism emerged from her experience as a housewife battling Hooker Chemical's Love Canal debacle asks, "How do you negotiate with someone who's killing your kids?"[43]

By serving as a matchmaker between environmentalists and corporations, the public relations industry has played an important role in promoting such false harmony. Firms like Hill & Knowlton have urged corporations to develop collaborative projects, make donations, advertise in green magazines, and secure seats on environmental groups' boards of directors. By doing so, they argue, corporations can muffle or quell future criticism by these groups.[44]

Examples of corporate America intertwining itself with mainstream U.S. environmental organizations are abundant. For instance, the World Resources Institute has more than thirty "corporate supporters," many of whom are major polluters. WRI, which also has a number of corporate leaders sitting on its board of directors, claims it does not "have the expertise" to judge the environmental impacts of transnationals that give it money. Nina Kogan, WRI's director of corporate relations, justifies its acceptance of money from Chevron, Citibank, DuPont, and WMX among others, because they're "the companies most actively trying to develop solutions."[45]

Similarly, the National Audubon Society, which says it is "very discriminatory" about who it accepts money from, lists WMX, Cargill, Chevron, Dow, DuPont, Ford, Motorola, Scott Paper, and a number of other environmentally destructive entities among the more than sixty major corporate donors that provide it with well over $1 million a year.[46]

Reclaiming the Blue Planet

Corporate globalization has become the dominant economic, and to some degree political, paradigm. Within this paradigm the phenomenon of corporate environmentalism has emerged, at once responding to the ecological contradictions of global corporate capitalism and attempting to appropriate the language, images, and objectives of the environmental movement.

Yet social movements continue to emerge to challenge the reign of corporate globalization, presenting both danger and opportunity. The danger is that a xenophobic nationalist reaction will arise as the foremost voice of opposition. The opportunity is that the nascent forces of grassroots globalization, which are building a series of networks across the Earth, have a chance to step forth with an alternative to the corporate planet.

Reclaiming the blue planet will require implementing mechanisms for democratic control over corporations and economies. Such democratization involves redefining both the concept of corporate accountability and the concept of the corporation itself. Corporate accountability is at present understood by many people as a corporation's nonbinding response to the demands of those affected by its activities. It is regularly defined as a transnational's responsibility to its investors, responsiveness to the demands of its workers or of a local community where it is operating, and as a company's voluntary reporting of environmental information. Indeed, corporate accountability is increasingly becoming confused with corporate self-regulation.

True corporate accountability, however, is much broader and more profound. It means that a company can be held strictly accountable to the laws and democratic processes of communities, governments, and the global framework in which it operates. At the heart of such democratic governance should lie the concept that corporations do not have any inherent right to exist, but rather are granted that right by "the people" and therefore must answer to the public. Thus the people, through a process of democratic political representation, should have the right to define "the corporation"—what it is and what it can and cannot do. If a company consistently violates the law, the public should be able to petition the government to revoke its charter and dismantle it by liquidating its assets.

With corporations under more democratic control, monopolies can be broken up. Economies could become focused more on sustainablity. Such a climate could foster a decentralized ecologically oriented market economy based on the development of clean technologies, organic agriculture, worker-controlled enterprises, small businesses, municipal corporations, alternative trade associations, banks that are truly dedicated to the eradication of poverty, and more. Building such a vision (one that would dismantle global capitalism as we know it) may seem a quixotic quest. Yet, many interim steps can be taken to build democratic control over corporations while fostering alternative economic development. The following section provides some of the broad outlines of what such a transformation might entail.

Local Democratization

Creating mechanisms that strengthen communities at the local level with rights, obligations, and information is essential for achieving democratic governance of corporations. As Vandana Shiva argues in her insightful essay "The Greening of Global Reach," while transnational corporations are parochial entities emanating from a very local area and projecting themselves globally, the converse is also true. Local communities, when taken as a group, are truly global, or "planetary" entities. While they are often isolated from one another, they share many characteristics, such as a spiritual and tangible connection with the Earth and her resources. This connection provides the foundations of a nearly universal ethic of sustainability that runs counter to the growth and profit-accumulation ethic of corporate capitalists. Thus, suggests Shiva, equipping local communities with the right to timely, accurate, and comprehensible information about corporate and governmental development projects, and with the right to ultimately reject such projects—in other words, democratizing decision-making processes at the local level—constitutes what she calls "a new global order for environmental care."[47]

Such empowerment could go a long way toward breaking up corporate control and decentralizing power, placing communities in control of their resources, ecosystems, economies, politics, cultures, and destinies. It could also alleviate many current social and environmental disasters and prevent many in the future.

How local democratization can feed into broader, larger-scale national and international political processes without being co-opted or corrupted is a difficult question. This forms part of the paradox and challenge of thinking and acting both locally and globally.

National Democratization

The pervasive corporate role in government, which permeates most nations, must be addressed. Such a change implies reinventing the nation-state in a fashion that transforms it from an elitist, unaccountable, overly bureaucratic, often corruption-ridden entity, to a new model of governance that is an expression of grassroots democracy.

However, the tide is moving swiftly in the opposite direction, spawning only eddies of resistance. As Indian scholar Rajni Kothari lucidly points out, the Right has co-opted the Left's critique of the nation-state and used it to promote its own "free market" agenda.

> While there is no doubt that too much dependence on the State had produced a structure of power and authority, at once insufficient and oppressive, as was indeed continuously pointed out by the left-cum-liberal critique of the State . . . today that very critique is being hijacked for a wholly opposite and right wing purpose by the proponents of globalization. . . . It has. . .become the basis for reinforcing privatization and debunking not just the public sector but the State as such, not just the bureaucratic excesses but democratic politics itself, not just corruption in public life but the public arena as a whole.[48]

Thus the criticisms of the overcentralizing, antidemocratic nature of the state, of too much "big government" are being used by corporate interests to attempt to dismantle as much of government as possible. Nevertheless, argues Kothari, the nation-state must be reinvented as an instrument of democracy and popular participation that promotes the interests of the poor, of labor, and of the environment.

Of course, this is nothing less than a revolutionary agenda. The transformation of the nation-state requires that powerful, broad-based movements arise and break the strong link between corporations and government. This could create the political space for democratically accountable legislators to shift government subsidies away from pockets of large corporations in the petroleum, nuclear, and timber industries, for instance, and toward support for the development of alternative, ecologically sound technologies and industries, as well as education, health, and social programs for the poor. It could facilitate the introduction of environmental accounting and government support for other mechanisms, such as green taxes, that promote sustainability. Such changes could alter regulatory structures so that the big transnationals would lose many of the unfair advantages they currently enjoy, thus permitting small businesses to thrive.

In a country like the United States, it could provide for the possiblity of stringent national charters for corporations, rather than the current state-based system, which has allowed half of the top 500 U.S. corporations to seek haven in Delaware, where regulations are the weakest.[49] It could encourage support for a constitutional amendment that would deprive corporations of their Supreme Court–granted legal status as individual persons. It could give government the mandate to ban particular production processes, products, or entire industries—such as nuclear or chlorine sectors—while implementing programs that would provide transitional support along with meaningful education and retraining for workers.[50]

Greater democracy at the national level in the northern industrialized countries could have a number of global repercussions as well. For instance, communities in countries such as Costa Rica, Colombia, Ecuador, or India (Bhopal), which have attempted to sue U.S. corporations such as Dow Chemical, Shell, and Texaco for destroying their lives and resources, would have a much better chance of gaining standing in U.S. courts of law—something that has, until now, been difficult to achieve. Parliaments, diets, and congresses less influenced by transnational corporations would be much more likely to extend national bans on deadly pesticides, hazardous pharmaceuticals, and products such as asbestos to the export of such lethal merchandise as well. These legislatures could also mandate that the corporations they charter follow the highest standards throughout their international operations. Conversely, if countries in the South and the East exerted more democratic control over corporations, they would be more readily able to apply social and environmental criteria to screen and control foreign investors.[51]

Global Democratization

Finally, true accountability means that international institutions such as the United Nations (UN) must be imbued with greater transparency and democracy themselves, while taking more significant steps to govern the behavior of transnational corporations at a worldwide level. Many regulations, standards, and market mechanisms must be global. For too long the transnationals have played governments against one another in a gambit to undermine workers' power and environmental laws. A system of global governance is necessary to control transnational corporations while raising labor and environmental standards. Such a role could be undertaken by the UN and financed by a modest tax on international financial transactions, as originally proposed by Nobel economist James

Tobin. The income generated by such a tax would at once lessen the UN's dependence on the world's richest governments and provide it with more resources than all international agencies currently have at their disposal.[52]

As a United Nations study points out, "despite large gaps and great disparities between the regions, there is now a significant body of rules of international environmental law at the regional and global levels. Those rules, in one way or another, affect every geographic and political region and state, and influence, directly or indirectly, virtually every type of industrial activity, including TNCs [transnational corporations]."[53] This body of law could be further strengthened, elaborated upon, and refined to address directly the unique, often stateless nature of the global corporations. For instance, as we move into the greenhouse age, it is possible and foreseeable that protocols will be attached to the climate convention that specifically regulate and control the activities of the corporate sectors most responsible for climate change. Similar protocols to the biodiversity treaty could be created to force greater accountability from fisheries, forestry, and biotechnology transnationals.[54] The United Nations is also working on an intergovernmental treaty to phase out and ban persistent organic chemicals such as chlorine.

The potential that such agreements hold, however, is seriously diluted, not only by corporate lobbying against them, but also by the dynamics of corporate globalization. The increasing power that is concentrated in institutions that define themselves as primarily economic in orientation, such as the World Trade Organization and World Bank, serves to weaken the more socially, environmentally, and politically oriented United Nations. This has profound political ramifications that must be addressed. Indeed, mechanisms for global corporate accountability that do not subordinate this de facto system of corporate governance to more democratic institutions oriented according to the prerogatives of ecological sustainability and social justice will ultimately fail.

The only way that these ideas about reclaiming the blue planet will be transformed into reality is through pressure provided by a global grassroots movement—pressure that uses campaigns, boycotts, strikes, and direct action to effect change. The road toward grassroots globalization will be arduous. Yet in many respects this journey follows a path that has been well traveled and chronicled by those seeking justice throughout the annals of civilization. This magnificent human adventure has taken many different political, cultural, technological, and moral twists and turns over the centuries, but the quest continues unabated. The path of sustainability, justice, and democracy has not disappeared. It can be found by reading the compass of the world's valiant history of struggles for social justice.

Notes

1. See Wolfgang Sachs, "The Blue Planet: An Ambiguous Modern Icon," *Ecologist* **24**(5) (September/October 1994):170–175.

2. These webs are cogently described throughout Richard J. Barnet and John Cavanagh, *Global Dreams: Imperial Corporations and the New World Order* (Simon & Schuster, New York, 1994).

3. Stephan Schmidheiny, *Changing Course: A Global Business Perspective on Development and the Environment* (Cambridge, Mass.: MIT Press, 1992, pp. 1–178).

4. Harris Gleckman with Riva Krut, "Transnational Corporations' Strategic Responses to 'Sustainable Development,'" paper prepared for *Green Globe 1995* (New York: Oxford Univ. Press, 1995, p. 3).

5. World Resources Institute (ed.), *Corporate Environmental Accounting* (Washington, D.C.: WRI, 1995).

6. See for instance, Schmidheiny, *Changing Course,* pp. 181–334; Jan-Olaf Willums and Ulrich Goluke, *From Ideas to Action: Business and Sustainable Development, The Greening of Enterprise 1992* (International Environmental Bureau of the International Chamber of Commerce, Norway, May 1992); Kurt Fischer and Johan Schot (eds.), *Environmental Strategies for Industry: International Perspectives on Research Needs and Policy Implications* (Washington, D.C.: Island Press, 1993); Charles S. Pearson (ed.), *Multinational Corporations, Environment and the Third World* (Durham, N.C.: World Resources Institute/ Duke Univ. Press, 1987); Bruce Smart (ed.), *Beyond Compliance: A New Industry View of the Environment* (Washington, D.C.: World Resources Institute, 1992).

7. ICC "Business Brief: Summary of the Series Nos. 1–9," prepared for UNCED by the International Chamber of Commerce (June 1992).

8. David C. Korten, "The Global Economy: Is Sustainability Possible?" presentation to the Presidio Conference, San Francisco (April 25, 1995).

9. Naomi Roht-Arriaza, "UNCED Undermined: Why Free Trade Won't Save the Planet," Greenpeace UNCED Reports, Greenpeace International (March 1992), pp. 9–10.

10. Testimony of Lori Wallach of Public Citizen's Congress Watch on the GATT Uruguay Round Agreement, before the Senate Commerce Committee (October 17, 1994).

11. Chakravarthi Raghavan, "South Urges More Study on TRIPs/Sustainable Development Interface," *Third World Economics* (July 1995):1–15; "Correcting Potential Conflicts Between MEAs and GATT," submission by Greenpeace International to the GATT's Commission on Trade and Environment (1994).

12. The PCSD was created by the Clinton administration in the aftermath of the Earth Summit in 1992 to bring industry and environmental leaders to the table with high-ranking government officials to work out a long-range plan that would move the United States toward greater ecological sustainability.

13. Michael Weisskopf and David Maraniss, "Forging an Alliance for Deregulation: Rep. Delay Makes Companies Full Partners in the Movement," *Washington Post*, March 12, 1995; Stephen Engelberg, "Business Leaves the Lobby and Sits at Congress's Table," *New York Times*, March 31, 1995; Stephen Engelberg, "Wood Products Company Helps Write a Law to Derail an EPA Inquiry," *New York Times*, April 26, 1995.

14. Natural Resources Defense Council, "State of Nature," Natural Resources Defense Council Legislative Watch (Washington, D.C., October 27, 1995).

15. John H. Cushman Jr., "Spending Bill Would Reverse Nation's Environment Policy," *New York Times*, September 22, 1995.

16. Natural Resources Defense Council, "Stealth Attack: Gutting Environmental Protection Through the Budget Process," NRDC on-line version (July 15, 1995): section 15.

17. Ibid, sections 1–14.

18. William K. Stevens, "GOP Seeks to Delay Ban on Chemical Harming Ozone," *New York Times*, September 21, 1995.

19. Marian Burros, "Congress Moving to Revamp Rules on Food Safety," *New York Times*, July 3, 1995.

20. Jane Fritsch, "Sometimes, Lobbyists Strive to Keep Public in the Dark," *New York Times*, March 19, 1996; NRDC, "Stealth Attack," section 3.

21. John H. Cushman Jr., "Congressional Republicans Take Aim at an Extensive List of Environmental Statutes," *New York Times*, February 22, 1995.

22. Alan Thien Durning, "Can't Live Without It," *World Watch* (May/June 1993):18.

23. Narrator's comment in The Prize, Part Seven, "The New Order of Oil," an Invision Production for Majestic Films and Transpacific Films in association with BBC TV and WGBH Boston.

24. Dow, "What on Earth is Dow Doing?" Dow Chemical Company advertisement.

25. Joshua Karliner, "God's Little Chopsticks," *Mother Jones* (September/October 1994):16.

26. Kenny Bruno, *The Greenpeace Book on Greenwash* (Washington, D.C.: Greenpeace International, 1992, p. 20).

27. ICI advertisement, *The Malay Mail*, April 5, 1993.

28. Kenny Bruno and Jed Greer, "Imperial Chemical Industries" Greenwash Snapshot #10, Greenpeace International (1993); Walden Bello, *People and Power in the Pacific* (San Francisco: Institute for Food and Development Policy, 1992, p. 57).

29. Lewis C. Winters, "Does it Pay to Advertise to Hostile Audiences with Corporate Advertising?" *Journal of Advertising Research* (June/July 1988):11–18.

30. Justin Lowe and Hillary Hansen, "A Look Behind the Advertising," *Earth Island Journal* (Winter 1990):27.

31. Greg Lyon, Target Four Investigation of Chevron Advertisements, KRON-TV news video, San Francisco, n.d.

32. Michael C. Lasky, "Earth-Friendly PR Claims Must Meet FTC's 'Green' Guidelines," *O'Dwyer's PR Services Report* (February 1995).

33. Jack Doyle, *Hold the Applause!: A Case Study of Corporate Environmentalism as Practiced at DuPont* (Washington, D.C.: Friends of the Earth, 1991, p. 53).

34. Ben Bagdikian, *The Media Monopoly*, 4th ed. (Boston: Beacon Press, 1992) cited in John Stauber and Sheldon Rampton, *Toxic Sludge Is Good for You* (Monroe, Maine: Common Courage Press, 1995, p. 181; N. John, "The Global Media: Caught in a Whirlpool of Bad Logic," Third World Network Features (1994).

35. Ronald K.L. Collins, *Dictating Content: How Advertising Pressure Can Corrupt a Free Press.* (Washington, D.C.: Center for the Study of Commercialism, 1992, pp. 27–30); William Greider, *Who Will Tell the People: The Betrayal of American Democracy* (New York: Simon & Schuster, 1992, p. 329).

36. Collins, *Dictating Content,* pp. 9–12; the $3 billion figure is from "Corporate Advertising Expenditures 1988–1992 in Nine Media," Leading National Advertisers/Arbitron Multi-Media Services, Publishers Information Bureau, New York (1993).

37. Stauber and Rampton, *Toxic Sludge Is Good for You,* p. 125.

38. Suzanne Alexander Ryan, "Companies Teach All Sorts of Lessons with Educational Tools They Give Away," *Wall Street Journal,* April 19, 1994; Jodi Mailander and Cyril T. Zanesky, "Big Business and the Classroom," *Miami Herald,* April 17, 1994; Stewart Allen, "Exxon's School Spill," *San Francisco Weekly,* December 9, 1992; Keith Schneider, "Exxon Is Ordered to Pay $5 Billion for Alaska Spill," *New York Times,* September 17, 1994.

39. John J. Fried, "Firms Gain Hold in Environmental Education," *Philadelphia Inquirer,* March 27, 1994.

40. David Lapp, "Private Gain, Public Loss," *Environmental Action* (Spring 1994); Ryan, "Companies Teach."

41. Anon, "Links with Activist Groups Get Results in Environmental PR," *O'Dwyers PR Services* 8(2) (February 1994):1, 20–22.

42. Jonathan Lash, remarks made during a PCSD press briefing, Washington, D.C. (October 26, 1994).

43. Lois Gibbs, author's interview (March 21, 1995).

44. Anon, "Links with Activist Groups," pp. 1, 20–22.

45. Nina Kogan, author's interview (March 14, 1995); "Corporate Relations Program," World Resources Institute brochure (1995).

46. Kimberlee A. McDonald, Director of Foundation Relations, National Audubon Society, author's interview (March 14, 1995); National Audubon Society annual report (1994), p. 13.

47. Vandana Shiva, "The Greening of the Global Reach," in *Third World Resurgence,* No. 14/15, (Penang, Malaysia) (1992):58.

48. Rajni Kothari, *Growing Amnesia: An Essay on Poverty and the Human Consciousness* (Viking Penguin India, New Delhi, 1993, pp. 9–10.

49. This figure is from the 1970s; Jonathan Rowe, "Reinventing the Corporation," *Washington Monthly* (April 1996):19.

50. Oil, Chemical, and Atomic Workers Union, "Understanding the Conflict Between Jobs and the Environment: A Preliminary Discussion of the Superfund for Workers Concept," Oil, Chemical and Atomic Workers International Union, pamphlet (1991).

51. See Kenny Bruno, "Screening Foreign Investments: An Environmental Guide for Policy Makers and NGOs" (Third World Network, Penang, Malaysia, 1994).

52. Sherle R. Schwenninger writes that "a modest levy of 0.1 percent would be so small as to have virtually no effect on capital flows but large enough . . . to yield $300 billion." "How to Save the World: The Case for a Global Flat Tax," *The Nation* (May 13, 1996):17.

53. TCMD, International Environmental Law: Emerging Trends and Implications for Transnational Corporations, Department of Economic and Social Development, Transnational Corporations and Management Division, Environment Series No. 3 (New York: United Nations, 1993):xv.

54. Harris Gleckman with Riva Krut, "Transnational Corporations' Strategic Responses to 'Sustainable Development'" in *Green Globe* 1995 (Oxford Univ. Press, 1995, p. 3.

11

Selling "Mother Earth": Advertising and the Myth of the Natural

Robin Andersen

A television advertisement for Kitchen Aid opens with spectacular vistas of purple mountains framing a desert landscape. A foregrounded spire juts heavenward, rising from scarred hills cut by centuries of water runoff. Mist hangs between the barren ranges. This image fades to an oven framed by kitchen counters decorated in the same sun-washed shades as a voice intones, "Kitchen Aid ranges mirror nature in surprising ways." The camera moves across another exquisite desert view. This time the camera is closer, the hills are smaller, and sparse desert flora cover the dry earth. The camera moves over ancient hills rounded by centuries of wind erosion; superimposed are the words "Kitchen Aid."

These desert contours morph into the visually similar curved forms of twisted bread baking inside the oven. The voice adds, "with even warmth." Next, as the camera glides high above the protruding formations, it moves toward us and passes over a sheer, free-standing rock face. The image transposes from the sun-washed rock wall to the off-white oven door, moving upward to close. The voice assures, "strength that endures." Next, "a mystery that unfolds," reveals, through time-lapse photography, cactus flowers opening. With the same pacing, the image turns into a rising soufflé. "The fire of creation" fills the screen with the desert sun's warm golden glow. The golden circle fades into the lit stovetop burner, then into a circular skillet sautéing vegetables, "all someplace a little closer to home. Kitchen Aid freestanding ranges, built-in ovens and cook tops," the ad continues.

Inside the kitchen now, a woman moves around the appliance-laden, picture-perfect room. She wears a long Earth Mother, printed gauze skirt, and as we hear, "because we took a cooking lesson from Mother

Nature," the image moves back to a panorama of the desert with "Kitchen Aid" slashed across the hot, dry landscape. Throughout the ad, the female voice, soft and low, mixes with continuous new-age spiritual humming that rises to an inevitable crescendo as the commercial ends.

There is no essential, authentic, or inherent connection between the desert and the oven. A pristine ecosystem bears little logical connection to a manufactured appliance. They are both hot, though not comparable in degree, but the sensuous analogy establishes a connection. More important, however, stunning visual juxtapositions and editing forge a powerful and compelling association that unites the natural world with the commodity. They become indistinguishable; our social constructions and commonsense conceptions of "untouched wilderness" are now compatible with the highly materialistic lifestyles of first-world consumer culture.

This chapter seeks to demonstrate how corporate consumer culture creates enchanting and persuasive advertising messages, using both visual and textual strategies, that celebrate the environment while their own business practices continue on a path of ecological destruction. Left unexamined, such influential messages help facilitate corporate environmental destruction, especially in the absence of information and public debate about the destructive consequences of consumer culture.

The Kitchen Aid ad inserts the oven into the untarnished landscape through the use of a highly persuasive mode of visual language. Visual persuasion uses symbolic, associative terms that lack "propositional syntax", that is, they do not explicitly indicate causality or other logical connections.[1] This syntactical indeterminacy is one important aspect of the persuasive uses of images of nature in advertising.

The symbolic landscapes of consumer culture, including visually enhanced images of the natural world, allow advertisers to make insinuations about their products without explicitly claiming anything. To state openly that an oven is like nature because it is hot, and because a rising soufflé can be made to look like the opening of a cactus flower, would sound ridiculous. However, because the visual implications have not been stated plainly, they are not logically rejected. Such stunning associations make powerful connections between the natural world and the world of products; pleasing aesthetic representations and feelings about awe-inspiring natural landscapes are united with the product. The sense of psychic pleasure and inspiration is linked to the Kitchen Aid freestanding ranges. Kitchen Aid used these techniques in a series of ads for their appliances.

Perhaps most significant is their refrigerator advertisement. The aerial camera glides though clear, blue skies over a high-mountain forest covered in pure, white snow. As we hear the words "crisp, freshness," the image cuts to a refrigerator vegetable compartment. And with "brilliant light," another snow-covered peak and a dazzling bright-white mountain meadow are transfigured into the refrigerator's interior. The voice proclaims, "The refrigerator designed with a blueprint from Mother Nature, by Kitchen Aid." The last shot is a high-mountain lake luminous with the reflection of snow-covered peaks, as a woman's voice whispers, "For the way it's made." But what *about* the way it's made?

Advertising's symbolic culture inserts the product into the natural world, removing it from the context of its production. The ad never explains how it is made, who made it, or the effect on the environment, either during its production or during its use. The remarkable juxtapositions silence those realities and hide the social and environmental relations of the production and use of the commodity.

However, there is much to say about the "way it's made." The primary industrial cause of ozone depletion is the gases associated with refrigeration. Chemicals such as chlorofluorocarbons and hydrochlorofluorocarbons (CFCs and HCFCs), widely used as refrigerants and insulating foams, destroy the ozone layer by releasing chlorine in the upper atmosphere. Refrigeration, however, no longer requires ozone-depleting gases. Over one million "greenfreezes" have been sold in Europe, but such environmentally friendly refrigerators are not available in the United States. A campaign by Greenpeace encourages American manufacturers to offer nonozone-depleting refrigerators to the American public. The campaign targeted Whirlpool, the parent company that manufactures Kitchen Aid, because it is a huge company that produces parts for the greenfreeze sold in Europe. Whirlpool actually sells compressors to some leading manufacturers of the greenfreeze. However, the attempts have failed. U.S. companies will not make them here.

Instead, Whirlpool chose to respond with a slick advertising campaign. Even as Whirlpool capitalized on persuasive portraits of natural beauty, it refused to invest in the new technology available for making environmentally safe refrigerators, creating a symbolic culture that seemingly reveres the environment as it helps destroy it. This public relations strategy has been referred to as "greenwashing"; it has become a cornerstone of American corporate promotional culture.

Instead of creating popular support by moving toward sustainable resource management, less toxic ingredients and manufacturing, waste reduction and reusable packaging, corporations rely on perception management through advertising as a major feature of their strategies. Perception management creates symbolic associations that evoke a sense of well-being. Since the advertising taps directly into psychic associations using the language of art, poetry, and the unconscious, the logical mind does not reject absurd representations about "taking a cooking lesson from Mother Nature," because as psychoanalysts Haineault and Roy argue, such impressions are psychologically pleasing.[2] Rejecting them would require a degree of distress the psyche seeks to avoid. The destructive qualities of the product and its negative environmental effects are successfully hidden. In an age when the public is bombarded with 1,500 advertising messages a day, and when corporate ownership and advertising influence successfully block information about environmental destruction by corporations, it is difficult to nourish citizen awareness and action.

Such false connections exist in a media environment that offers little information to counter advertisers' representations. With a few notable exceptions, major advertising clients have been successful in putting direct pressure on editors and producers not to contradict the messages of their advertising campaigns in programming content.[3] The only way the impressions created by advertising culture can persist is in an atmosphere devoid of contradiction.

There is another aspect of advertising's appropriation of nature that also makes it a powerful tool for greenwashing and, more broadly, an enchanting and influential celebration of capitalism as a social system. In addition to the visual persuasions and the lack of information are the underlying belief systems and commonsense conceptions of nature that much advertising successfully mines. How can we accept the assertion that Whirlpool "took a cooking lesson from Mother Nature," or that ideas about the industrial design of appliance manufacturing "flowed from Mother Nature"? Why do these and other absurd presentations, even without direct information on ozone depletion, sound acceptable to a public willing to revere nature but at the same time participate fully in consumer culture?

Media Representations and Cultural Conceptions of Nature

In large measure, the visual constructions of nature in advertising can be linked to, and have been appropriated from, some key conceptions of the environmental movement. The advertisements are effective because they do not contradict the "commonsense" understandings of the natural world shared by many. These compatible portrayals, rooted in ideological assumptions about nature, must be examined and redrawn in order to foster public awareness and citizen action and achieve sustainable social and environmental ecosystems.

Many environmental writers regard the space missions as key cultural markers in our contemporary conception of nature, the Earth, and the need for conservation: "It is almost commonplace to note how the first pictures of Earth from Space in 1966 made evident the frailty of the planet and sparked a global ecological consciousness."[4] Ecofeminist Chaia Heller (1993) writes, "Awareness of the ecological crisis peaked in 1972 when the astronauts first photographed the planet, showing thick furrows of smog scattered over the beautiful green ball. 'The planet is dying' became the common cry. Suddenly the planet, personified as Mother Earth, captured national, sentimental attention."[5] The mass media presented pictures of the Earth as well as representations of the significant environmental issues of the last 30 years, from protecting endangered species around the globe to preventing the destruction of the rain forests and saving the whales.

Indeed, even though most people have not been to the moon, or the faraway wild places they become concerned with, television creates critical impressions. As DeLuca observes, "in particular PBS documentaries, I learn of and become concerned about the Amazon rainforests."[6] However, television and other media representations are just that, (re)presentations. They are stories of nature, not the Earth made real. We know Mother Nature through the stories we tell about her. Our cultural narratives help shape our perceptions.[7] Ecosystems, wilderness areas, and wild things are most often unknown to us without the framing of such cultural texts. And those texts will affect not only our perceptions of nature but also the way we act in and upon the natural world.

Heller argues that we tell a particular type of cultural narrative about Mother Nature. This narrative features characters and relationships borrowed from the medieval romanticization of women as depicted in passionate love poetry: "The metaphors and myths of this eco-drama are plagiarized from volumes of romantic literature written about women, now recycled into metaphors used to idealize nature."[8] The medieval narrative drama of romantic love plays out in the wistful longings of a man for an idealized, pure woman whom he vows to protect, for his love can never be consummated. Rooted in Platonic dualism, the ideal love is unpolluted by physical contact. She is powerless and in need of protecting. The lover realizes his romantic fantasy only through noble self-control and heroic acts of protection and sexual self-restraint. The beloved is deserving because she is pure and chaste.

Heller argues that the imagery of the contemporary ecology movement finds its ideological roots in this cult of the romantic. Mother Nature takes the role of the helpless beloved, and those in the ecology movement vow not to defile her. "In our modern iconography, nature became rendered as a victimized woman, a Madonna-like angel to be idealized, protected, and saved from society's inability to constrain itself."[9]

Procter & Gamble's advertising campaigns display similar iconography and the same sentimentalized message. In soft focus, a little girl drinks water from a beautiful indoor sink. In the next picture a deer drinks from a pond in an idealized outdoor setting. The caption reads, "everybody deserves a clean home." This series also includes images of a baby girl playing with a little yellow duck, while fuzzy little yellow ducks swim in a placid soft-focus lake in the picture next to her. Helpless and idealized, Mother Nature must be protected from all those who would destroy her.

The sentimentalized renderings of romantic ecology raise many questions. For example, whom does Mother Nature need to be protected from? This category is always vague and sweeping, but the implied formulation is often "human nature," as with Devall and Session's formulation in *Deep Ecology*:

Excessive human intervention in natural processes has led other species to near-extinction. For deep ecologists the balance has long been tipped in favor of humans. Now we must shift the balance back to protect the habitat of other species.[10]

Condemning all of humanity as equally responsible for environmental degradation ignores oppressive social and economic relations, as well as the unequal distribution and use of natural resources and wealth. Certainly "failing to expose the social hierarchies within the category of human erases the dignity and struggle of those reduced to and degraded along with nature."[11] The exploited laborers of the third world who toil in extreme conditions in sweatshops making commodities for first world consumers are not equivalent to the ravages of global capitalism.

Overpopulation must certainly be a concern, not only for environmental organizations, but for all of the globe's peoples. Yet some would put the central responsibility for the problem squarely on the shoulders of women—human mothers defiling an idealized Earth Mother. While women own and control only about 1 percent of the Earth's wealth, they account for 80 percent of human labor power. United Nations population studies demonstrate that in countries where women have become less economically exploited and more educated and socially empowered, the birth rate has declined.

The first world consumes almost four times more of the Earth's resources than its third world counterparts. Those who are victimized by and struggle against the ravages of global capitalism are not to blame for the ecological crisis, corporate global practices are. Habitually emphasizing birthrates out of context only facilitates capitalism's ravages of the Earth.

The idea that every person is equally culpable of defiling the Earth conceals the role of corporations in ecological degradation and distracts attention from the production, use, and disposal of resources. Public service announcements and advertising campaigns often suggest that solutions to environmental problems are the private responsibility of individual consumers. "We recycle" and recycling logos on many products are omnipresent, but such campaigns help manufacturers prevent mandatory deposit and recycling regulations. As long as manufacturing does not make recycling, especially of plastics, more effective, recycling will have little positive impact. The romantic narrative focus on individual responsibility and such socially constructed stories place corporate responsibility outside their discourse.

In this cultural ecodrama that we present, and read ourselves into, how are the characters drawn? What is the nature of this relationship between the beloved Mother Nature, and the protective ecoknight? As Heller points out, the romanticization of nature is "based on the lover's desires, rather than on the identity and desires of the beloved." The role of nature as the beloved is to offer emotional well-being to dedicated environmentalists and consumers alike. Some "daily affirmations" of new-age environmentalism begin to sound very similar to advertising messages. "I hold in my mind a picture of perfection for Mother Earth. I know this perfect picture creates positive energy from my thought, which allows my vision to be manifest in the world."[12]

An advertisement for Evian water mirrors these sentiments. Evian's campaign features healthy people delighting in pristine natural settings. On the left half of a two-page ad, a man runs across a blurred, cold, dark-blue background. Across his chest are the words, "In me lives a wildcat who chases the moon and races the wind and who has never measured his life in quarterly earnings." The opposite page is an image of the French Alps, rendered in soft beige tints. The perspective places us on one of the mountains, peering over, but within, the snow-covered, jagged peaks. The caption claims, "In me lives the heart of the French Alps. Pure, natural spring water perfected by nature. Untouched by man, it is perfect for a wildcat." On the left page the word "wildcat" stands out in boldface, and on the right page "heart" appears in boldface. This graphic design unites the two pages; at a glance the viewer sees **wildcat heart**.

As with many ecologically posed ads that feature the natural world, this one constructs an alternative sense of place, a pure, untouched environment offered to the ecoknight turned consumer as an escape from the unpleasant realities of modern industrial life. The goodness of Mother Nature is generally depicted in advertising by what she can do for us and most often by the way she can make us feel. Turned into a commodity, nature promises a return to a state of emotional well-being from the strain and stress of urban and social life.[13]

In addition, the absence of a social presence in the renderings of wild places reinforces a solitary, individual relationship to nature that negates collective action. In the Evian ad, the heart of Mother Nature nurtures the wildcat, the human spirit supernaturally transformed and empowered. This is what Mother Nature can do for us. Such ideas are compatible with a new-age environmentalism based in the romantic tradition. However, once turned into a commodity through advertising, romantic attainment can only be realized through the possession of the product tied to those sentiments. With all those wildcat spirits consuming water sold in plastic bottles, what happens to the environment?

The Costs of Consumption

Over the last decade the marketing of bottled water has vastly increased sales of the product. Instead of relying on municipal water supplies, especially when away from home, it has become popular to buy and carry individual water bottles made of plastic. Plastic is made from nonrenewable fossil fuels and manufactured from fractions of crude oil or natural gas changed into solid form through the use of different chemicals. Many of these chemicals are highly toxic, such as benzene, cadmium and lead compounds, carbon tetrachloride, and chromium oxide. They are used as solvents, for coloring, and as catalysts in chemical manufacturing. Over the past two decades, the production of synthetic chemicals has almost tripled, and much of this increase is attributed to the manufacture of plastics. Synthetic chemicals released into the environment create dioxins and other toxins.

Thermoplastics, the soft type of plastic, such as polyethylene, are used for everything from milk jugs and margarine tubs to pipes and tubing. Polystyrene is used to make styrofoam, other packaging material, and tiles. The most toxic plastic is polyvinyl chloride (PVC), and it is used for pipes, shower curtains, and insulation, and as Greenpeace points out, many children's toys and accessories. Polyethylene terephthalate (PET) is used to make plastic bottles.

According to the U.S. Environmental Protection Agency's (EPA) home pages on the World Wide Web, in 1994 only 4.7 percent of all plastic was recycled in the United States. Recycling is difficult because of the mix of plastics that cannot be degraded together. Incineration reduces the plastic headed for the landfill, but it also releases toxic pollutants, including heavy metals, into the air and through the remaining ash. These pollutants also enter the groundwater and contaminate the water supply. The EPA recommends what is called "source reduction" for plastics, to reduce the amount of plastic purchased and thrown away.

Instead of reducing the use of plastic, however, the marketing and distribution of bottled water has led to a dramatic increase in its use. The development of polyethylene terephthalate (PET), the lightweight, durable yet malleable plastic, allowed the manufacture and distribution of smaller, individual containers which continue to proliferate. Portable plastic, huge advertising budgets, and a profit margin of 40 to 45 percent (compared with 30 to 35 percent on soda) have led to an increase in the sales of bottled water by an amazing 144 percent over the past decade.[14]

Thus we have come full circle. The concept Mother Nature helps associate bottled water with purity and inspiration and compels people to express their love for nature by buying individual plastic containers, a substance that degrades the environment, presents hazards to public health, and ultimately has the potential to contaminate our drinking water.

Such uses of mythic nature help cloud judgment, as well as the knowledge and understanding of how consumer culture actually affects the environment. Nature is perfect purity, undefiled, "untouched by man," waiting there, after centuries, just to quench the spiritual thirst of the running man in the Evian ad. As with many advertisements, nature exists to provide solace and satisfy the spiritual longing of those who love her. Like all myths used to reinforce ideologies, this idealization of pure, untouched nature has been drained of any essential meaning. Nature exists only as it is reflected in the needs and desires of the Evian runner.

Loss of any true knowledge of the beloved is another consequence of grounding our love of nature in romantic mythology. "The romantic's love depends on his fantasy of his beloved as inherently powerless and good as he defines good."[15] This type of "knowing" nature is unidimensional and is "wedded to ignorance. Certainly the romantic does not know his lady to be a woman capable of self-determination and resistance."[16] Just as the true qualities of the beloved are never known to the lover in romantic poetry, when romanticized, nature itself remains a mystery. As long as nature is made myth, is docile and willing to give, she is deserving of love and protection.

However, this is a selfish love indeed. For once Mother Nature challenges the ego, refuses the needs, or engages in self-expression and determination, anything can be done to her. These points are borne out in the most common representations of nature in advertising today, those promoting sport utility vehicles (SUVs).[17] While the male would-be drivers are invited to experience the wild world of jungles, forests, deserts, and even underwater ecosystems to gain a sense of inspiration and adventure (again, nature making the hero feel good), nature can also be unruly and anything but docile.

Nature is often depicted acting up, creating bad weather and rough roads. She will burn you, she will freeze you, she will try to blow you away, as one 4×4 ad threatens. This invites the worst kind of treatment, and nature must be tamed. SUVs are most often depicted off-roading; Mother Earth flies out from under the tires that are tearing up wildlands, disturbing wildlife, and breaking down river banks as the 5,000-pound vehicles cleave through them. Domination and mastery over nature is the result when she steps out of line.

However, simply portraying nature as female is enough to invite a certain behavior. In a Wrangler ad for jeans, a man is given a pink cake by his wife. In the next scene, he is shown smashing through the now huge pink cake in a forest, expressing a distinctive adolescent male rebellion.

The romantic gendering of nature has resulted in complicated and contradictory advertising messages that are used for very destructive purposes. Originally, the concept of Mother Nature was borrowed from Native American philosophy and other indigenous cultures. While it may have had a noble purpose, when this concept is lifted from its social and cultural contexts and reinserted within patriarchal capitalist culture, it is turned into a commodity and commercialized in ways that ironically come to support the economic practices that misappropriate the Earth's resources. This new hybrid of Mother Earth is now a woman destined to be dominated in the same ways that women are subjugated by the legacy of patriarchal capitalist culture.[18] In such a culture, protective paternalism quickly loses its tolerance when those subjugated—women, minorities, or those outside the dominant social order—express rage at their unequal treatment or make demands for self-determination.

This is how we come to such contradictory uses and depictions of nature in consumer culture. For example, studies show that those who purchase SUVs think of themselves as environmentally conscious. They want to enjoy the great outdoors and have adopted a vehicular consumer style that expresses that. At the same time, the ads invite them to dominate nature. Because nature has been portrayed as an unruly female, these two messages are not seen as contradictory.

In addition, a mythicized conception of nature, existing only to please those who admire her, does not invite knowledge and understanding of how humans might best interact with nature. Because of the ways we formulate our love for nature, environmental consciousness can sit comfortably in the cab of an SUV, a machine that uses far more gas than a smaller car. Extracting the greater amounts of fossil fuel needed to run SUVs helps destroy the wilderness areas and animal habitats, such as the last Alaskan wilderness, so lovingly depicted in the persuasive ads for these vehicles. SUVs increase greenhouse gases and therefore global warming because of their lower emission standards. Because of lobbying by the automobile industry, this situation is not likely to change. The same industry that professes a love of nature fights fuel efficiency, emission standards, and public spending on mass transportation while it continues to make as much as $10,000 gross profit on the biggest and most destructive SUVs.[19] This is greenwashing at its worst.

The Beauty of Nature and the Cosmetics Industry

If we turn to another major aspect of consumer culture, that of the beauty and cosmetics industry, we find that the advertising of such products often features the globe, the concept or language of Mother Earth, and the natural world. It is common for beauty and hygiene product advertising to extol "natural" ingredients, proclaim the benefits of nature, and in general, associate the Earth's goodness with a wide range of cosmetics and their ingredients. In cologne advertisements, women sit in fields of sun-drenched flowers, and in hygiene ads they are compelled to douche with fresh, flower-scented liquids. Advertisements feature the goodness of Mother Nature in striking contrast to the depiction of real women's experience of nature.

In a comical essay reprinted in the *Utne Reader* titled, "Forever Fresh: Lost in the Land of Feminine Hygiene," journalist Alison Walsh writes that she was "astonished to discover that most daytime TV commercials have one clear message: women leak, dribble, and smell. . . . Apparently women must buff, douche, diet, gargle, and primp constantly if they are to overcome their basic vileness." Walsh, noting a certain gender imbalance observes that if we are to judge from television commercials aimed at men, "Evidently men are just fine the way they are. They have a small problem with weight gain and graying hair, but mainly they are handsome, playful and successful." She ends her piece with a query as to why there is no masculine hygiene aisle in the drugstore.[20]

However, women have been told, "how not to offend," by the cosmetics industry for a long time. In an essay with the same title, Marshal McLuhan noticed an anxiety-producing message from Lysol in the 1940s. The magazine advertisement features a woman waist deep in swirling water as the words "doubt, inhibitions, ignorance, and misgivings" seem to be pulling her down. Distressed, with her arms raised helplessly in the air, the ad rebukes, "Too late to cry out in anguish. Beware of the one intimate neglect that can engulf you in marital grief. For complete Feminine Hygiene rely on Lysol."[21] Advertising, as McLuhan comments, continues to remind us of "the terrible penalties . . . that life hands out to those who are neglectful . . . left in sordid isolation because they 'offend.'" When a lovely woman "stoops to BO, she is a Medusa freezing every male within sniff. On the other hand, when scrubbed, deloused, germ-free, and depilatorized, when doused with synthetic odors and chemicals, then she is lovely to love."[22] For McLuhan, "the cult of hygiene and the puritan mechanisms of modern applied science" were all part of technological progress. "Fear of the human touch and hatred of the human smell are part of this landscape of clinical white-coated officials and germ and odor proof laboratories."[23] Feminist writers (Gaard 1993) have long noted that the domination of women's bodies coincides historically with the domination of the natural world. Today, the white-coated hygienists of the 1940s and '50s have been replaced with images of Mother Earth, but the message is the same.

Just as the domination of nature often results in its destruction, the advertising imperative that the female body be sanitized, tamed, powdered, and redolent only of perfumes has led to dire health consequences. In "Taking a Powder," journalist Joel Bleifuss documents epidemiological evidence connecting the use of talcum powder with ovarian cancer, the fourth leading cause of death in American women.[24] While talc and other toxins found in cosmetic products pose a serious health risk to American consumers, especially women, the industry is not regulated. A U.S. Food and Drug Administration (FDA) document found on their WWW home pages states, "a cosmetic manufacturer may use any ingredient or raw material and market the final product without government approval." The FDA does prohibit seven known toxins such as hexachlorophene, chloroform, and mercury, but for the remaining 8,000 ingredients used in the manufacture of cosmetics, the industry regulates itself. While the FDA has the power to pull dangerous products off the shelf, it rarely does so, "despite mounting evidence that some of the most common cosmetic ingredients may double as deadly carcinogens."[25]

It is impossible to assess advertising claims of "natural" ingredients of cosmetic products most of the time, but cosmetic products "are often contaminated with carcinogenic byproducts, or contain substances that regularly react to form potent carcinogens during storage and use."[26] Identified as one of the most dangerous toxins by FDA doctors and cancer researchers are nitrosamines, a group of carcinogens found in a variety of products from shampoos to sunscreens. One of these nitrosamines, N-nitrosodiethanolamine (NDELA), forms when some common cosmetic ingredients, such as triethanolamine (TEA) and Cocamide diethanolamine (DEA), interact with the nitrites that are used to preserve many products. Vidal Sassoon shampoo, for example, contains the toxin, Cocamide DEA.

Joel Bleifuss also uncovered a number of research reports documenting the toxins contained in cosmetic products.[27] In 1992, tests conducted by the FDA revealed a product that contained NDELA at a concentration of 2,960 parts per billion, but the agency will not publish the brand name.

The European Union does not allow more than 50 parts per billion of nitrosamine-producing chemicals in cosmetic products. In 1992, all 14 products tested that year by the FDA were contaminated with the nitrosamine carcinogens. Individual FDA scientists are speaking out. Drs. Harvey and Chou assert that with the "information and technology currently available to cosmetic manufacturers, N-nitrosamine levels can and should be further reduced in cosmetic products. "A social goal should be to keep "human exposure to nitrosamines to the lowest level technologically feasible by reducing levels in all personal care products."[28]

Even as Redkin hair product features a picture of the globe, contrasting heaven (beautiful hair) and Earth (an image of the globe), plastic bottles appear in the lower right corner of the ad. The pervasive plastic packaging used for cosmetic products (with the exception of those made by the Body Shop) usually end up in the landfill (as noted above). In addition, one common general-purpose plasticizer, adipate, or DEHA, used in processing polyvinyl and other polymers, is also used as a solvent or plasticizer in such cosmetics as bath oils, eye shadow, cologne, foundations, rouge, blusher, nail polish remover, and moisturizers. It can contaminate foods wrapped in plastic films. This common plasticizer contaminates groundwater through fly ash from municipal waste incineration and wastewater effluents from treatment and manufacturing plants.

The $20 billion plus a year cosmetics industry relies on a lack of media scrutiny, which was made apparent in 1998 when epidemiological studies and the carcinogenic contents of some cosmetics were identified as important censored news stories for that year.[29] "Few publications put effort into investigating the cosmetics industry, which is not surprising since the industry is a major magazine and newspaper advertiser. This is especially true of the women's magazines. Consequently, there is almost no coverage of the industry."[30]

What is necessary to unveil the toxic substances and environmental destruction hidden behind advertising's compelling symbolic culture? Corporate cultural conceptions of nature must be recognized and transformed. Romantic protection of a mythic version of nature must become obsolete because of the ease with which its appropriation serves the interests of economic exploitation and environmental destruction. Instead, we should strive "to know and care for the resistance of all living things that dwell in poisoned eco-communities, offering ourselves as allies in resistance to social and ecological degradation."[31]

Authentic love is based on knowledge not myth, and "allied resistance" enjoins citizens to offer their support for the struggles against the global corporate domination that sustains the production and use of toxic substances. Given the power of global corporations, we need to identify and support innovative forms of resistance in our homes around the planet with a unified strategy against industries that pollute the environment, oppress their workers, and promote toxic substances. Such solidarity would demonstrate an authentic love for nature, rejecting the image of drinking water bottled in plastic, or smelling "naturally" fresh with a toxic cosmetic product, or driving an SUV as having any relation to a healthy life or preservation of the ecosystem.

Notes

1. Paul Messaris, *Visual Persuasion: The Role of Images in Advertising* (Thousand Oaks, Calif.: Sage Publications, 1997).

2. Doris-Louis Haineault and Jean-Yves Roy, *Unconscious for Sale: Advertising, Psychoanalysis and the Public* (Minneapolis: Univ. of Minnesota Press, 1993).

3. Robin Andersen, *Consumer Culture and TV Programming* (Boulder, Colo.: Westview Press, 1995); Michael Jacobson and Laurie Ann Mazur, *Marketing Madness: A Survival Guide for a Consumer Society* (Boulder, Colo.: Westview Press, 1995); Matthew McAllister, *The Commercialization of American Culture: New Advertising, Control and Democracy* (Thousand Oaks, Calif.: Sage Publications, 1996).

4. Kevin DeLuca, "Constituting Nature Anew Through Judgment: The Possibilities of Media," in *Earthtalk: Communication Empowerment for Environmental Action,* Star A. Muir and Thomas Veenendall, eds. (Westport, Conn.: Praeger, 1996, p. 60).

5. Chaia Heller, "For the Love of Nature: Ecology and the Cult of the Romantic," in *Ecofeminism: Women, Animals and Nature,* Greta Gaard, ed. (Philadelphia: Temple Univ. Press, 1993 p. 219).

6. DeLuca, "Constituting Nature."

7. Ibid.

8. Heller, "Love of Nature."

9. Ibid.

10. Ibid., p. 223.

11. Heller, "Love of Nature," p. 226.

12. Ibid., p. 223.

13. See Paul Messaris, "Pristine, Damaged, and Nightmare Landscapes: Visual Aesthetics of American Environmentalist Imagery," unpublished paper, presented at the Seventh Annual Visual Communication Conference, Jackson Wyo. (June, 1993).

14. See the cover story by Corby Kummer, "What's in the Water?" in *New York Times Magazine,* August 30, 1998:41. Ironically, "standards set for municipal drinking water supplies are mandatory and are monitored and tested more often than for bottled water," while the bottled water industry has been self-regulated for years (p. 41). In addition, a good portion of the bottled water sold in the United States is simply filtered or deionized.

15. Heller, "Love of Nature," p. 222.

16. Ibid.

17. Robin Andersen, "Road to Ruin! The Cultural Mythology of SUVs," in *Critical Studies in Media Commercialism,* Robin Andersen and Lance Strate, eds. (London: Oxford Univ. Press, 1999).

18. For a discussion of a related position on "cultural essentialism," see Laura Pulido, "Ecological Legitimacy and Cultural Essentialism: Hispano Grazing in the Southwest," in *The Struggle for Ecological Democracy: Environmental Justice Movements in the United States,* Daniel Faber, ed. (New York: Guilford Press, 1998).

19. See Andersen, 1999.

20. Alison Walsh, "Forever Fresh: Lost in the Land of Feminine Hygiene," *Utne Reader* (September/October, 1996):32.

21. Marshall McLuhan, *The Mechanical Bride: Folklore of Industrial Man* (New York: Basic Books, 1967, p. 61).

22. Ibid.

23. Ibid., p. 62.

24. Peter Philips (ed.), *Censored 1998: The News that Didn't Make the News* (New York: Seven Stories Press, 1998).

25. Ibid., p. 30.

26. Ibid.

27. Ibid.

28. Ibid., p. 324.

29. Ibid.

30. Ibid., p. 31.

31. Heller, "Love of Nature," p. 235.

12

Green Living in a Toxic World: The Pitfalls and Promises of Everyday Environmentalism

Marcy Darnovsky

More Americans now recycle than vote for president. Guides to green lifestyles crowd the bookstores. Products, ads, and public relations campaigns target the "green consumer." Each trip to the market is punctuated by the question, "paper or plastic?"

Green practices such as these are all but ubiquitous in the United States today. For most people they are casual and sporadic. For a few—a small minority but one whose numbers have soared since the early 1990s—they are part of more deliberate ways of life known as "simple living" or "voluntary simplicity."

This simple living revival began with small meetings in church basements and community centers, and found its way within a few years onto Good Morning America and the Oprah Winfrey Show. By mid-decade, a professional trend watcher had placed "voluntary simplicity" and "involuntary simplicity" among the U.S. trends of the 1990s. Grassroots "simplicity circles" and "eco-teams" emerged, along with systematic efforts toward "building the voluntary simplicity movement."[1]

Taken together, the practices that range from casual green habits to voluntary simplicity activism can be considered an *environmentalism of everyday life*. Now widespread in American culture, everyday environmentalism is practiced by millions of ordinary people who conserve energy, shop according to criteria of environmental health and responsibility, minimize their use of pesticides and packaging, and otherwise green their households and their lives. Some of these green habits of daily living, especially recycling, have become thoroughly mainstream and uncomplicated. Others, such as buying organic food, require carefully weighing substantially higher costs and inconvenience against anxieties about pesticide residues and support for sustainable agriculture.

For better or for worse, everyday environmentalism now serves as a lens through which many Americans understand—and misunderstand—the environmental crisis. In part because compelling ways to "be an environmentalist" are scarce, in part because of the tattered appeal of political engagement, everyday environmentalism often becomes the standard of measurement for green commitment and responsibility.

I have adopted the phrase "everyday environmentalism" in order attempt a fresh look at its political meanings and potentials—an enterprise nearly unimaginable if it were tied to the pop sociology and marketing term "green lifestyle." This term is not meant to downplay the obvious political shortcomings of everyday environmentalism: its affinities with ideologies of individualism, with the green appeals of marketers and advertisers, and with the ongoing downsizing of economic expectations for all but the very wealthy. Many critics, enumerating and examining these shortcomings, conclude that everyday environmentalism is a dead end. I argue otherwise. Though many who recycle and carpool will not be interested in political understanding and involvement, I believe that green activists can use everyday environmentalism as a strategic point of departure for connecting environmental concern to broader social analysis, and for transforming this popular green sensibility into an *environmental imagination* that is an active political force.

The hopes and fears that motivate everyday environmentalism are a potent brew. Three-quarters of North Americans identify themselves as environmentalists, and concern about environmental problems is remarkably persistent—so much so that pollsters, political analysts, and anti-environmentalists have repeatedly been taken by surprise. Surveys also register broad discomfort about "materialism" and "greed" in American life,[2] and dissatisfaction with what Juliet Schor calls the "work-and-spend squirrel cage."[3] Suspicions about environmental toxins in food, water, and consumer products grow as the quality and accessibility of medical care decay. Parents worry about the habits and values their children absorb from commercial culture. Moral qualms about increasing inequality in American society both reinforce and compete with economic anxieties that reach farther up the class ladder than they have for many decades.

A social movement that can give a dissenting voice to these anxieties, and a radical shape to this mix of issues, has culture-transforming potential. Some green activists and thinkers have set out to build such a movement, seeking to join everyday environmentalism to the concept of "sustainable consumption." Though the politics of the emerging sustainable consumption movement are still in flux, it could well serve to push everyday environmentalism out of the household and supermarket and into arenas of activism and political engagement.

The Ambiguities of Everyday Environmentalism

Everyday environmentalism is simultaneously a private response to a social problem, a product of corporate marketers eager to tame the environmental imagination, and a remarkable achievement of green activism. These mongrel origins bequeath it a murky and unstable set of political and cultural meanings. Everyday environmentalism is fitful; inevitably inconsistent; and terribly vulnerable to corporate marketing, advertising, and public relations. Its practitioners are too often caught in the narrow confines of lifestyle politics and too often fail to perceive consumption as a social activity that cannot be separated from systems of production and power.

Everyday environmentalism also shares the tendency of other kinds of "personal politics" to devolve into varieties of individualism, and so may wind up fostering a retreat into green privatism and apathy. It may become an "enclave environmentalism" accessible only to the wealthy, many of whom imagine that they can build personal bubbles—or gated communities—of environmental purity. Even among the less affluent, everyday environmentalism may bind its practitioners ever more tightly to their somewhat greener homes, their consciences assuaged and their willingness to tackle broader and more difficult efforts sated.

However, everyday environmentalism shares the strengths as well as the weaknesses of personal politics. Much of its promise lies in its ability to articulate a *relationship* rather than an *opposition* between daily life and social structure, between individual practices and collective political responsibility. Like the voluntary simplicity movement of the late 1960s and early 1970s, everyday environmentalism speaks to yearnings for a way of life in accord with environmental, moral, and political beliefs. Like the consciousness-raising groups of early second-wave feminism, everyday environmentalism can link individual and household habits to the infrastructure of consumption that restricts what we buy, what we eat, where we live, and how we get to work or school. Everyday environmentalism can insert familiar commodities and daily activities into a network of global social relations: production practices and technological choices, labor conditions and distributional inequities, ecological impacts and the powers of global corporations. And it is prefigurative: It enables people to start, right now, to enact a vision of a green social order that gives high priority to sustainability and environmental equity.

Everyday environmentalism can also contribute to a politics of environmental change that acknowledges the importance of women's work. It can make visible the political and economic import of "keeping house"— that sphere of socially necessary work, that Marxist feminists of the 1970s called "reproductive labor." Acknowledging everyday environmentalism as both political practice and economic activity aids the feminist project of admitting "women's work" into the realm of political significance.

The View from the Left

Many left-leaning environmentalists—that is, greens who understand that environmental problems are embedded in capitalist social structures and power relations—have been decidedly unimpressed with everyday environmentalism. These radical critics usually acknowledge the various forms of everyday environmentalism only to lament or lambaste them. They disparage efforts to live and shop ecologically as ineffective and self-delusionary gestures, and often label simple living and downshifting as examples of an insidious individualism that precludes collective opposition.

"Most people in America today see themselves as environmentalists," says the jacket copy of Tom Athanasiou's *Divided Planet,* a popular account of the social and economic inequities that accompany and define the global ecological crisis. "They recycle their trash, drive a bit less, and shop for energy-efficient products. They think they're making a difference—but they're wrong. The real threats to our environment . . . can't be halted or even slowed by the feel-good environmentalism of the industrialized world."[4]

Similarly, in *Earth for Sale: Reclaiming Ecology in the Age of Corporate Greenwash*, long-time social ecologist Brian Tokar identifies green consumerism as one of the major factors in "the increasing political disengagement of the environmentally concerned public."[5] Timothy Luke, a professor of political science and, like Athanasiou, a contributor to the journal *Capitalism Nature Socialism,* argues that green consumerism (represented in this comment by the *150 Simple Things You Can Do to Save the Earth* series of the late 1980s and early 1990s) reinterprets the "whole ecological crisis . . . as a series of bad household and/or personal buying decisions."[6]

Other radical observers brush off simple living with equal certainty. "Voluntary simplicity is a way of co-opting people who are concerned about consumption," writes Gene Coyle, a thoughtful energy analyst, in a posting to an online conference on consumption and the environment. "It is the response of people who really want consumption to continue to rise, not the response which will result in any fundamental change in the culture."[7]

These left critics, who discern in everyday environmentalism cause only for hostility and despair, have astutely identified many of its pitfalls. Their points need not only to be carefully considered but extended and elaborated as well. However, as a strategy for responding to the green exertions of millions of people, the approach they suggest is politically sterile. Rather than belittle efforts toward greener habits and households, why not nurture the worthy impulses that fuel them? Instead of abandoning the recyclers, carpoolers, green shoppers, and downshifters to corporate greenwashers, would it not be far better to propel them toward social analysis and political involvement?

This will not be an easy task, nor is success by any means sure. I am concerned, however, that many of those best suited to undertake it seem blind to its importance and potential. Left-leaning greens have long articulated the connections between environmental degradation, corporate power, and capitalist systems of production and domination, and their efforts would be key to deepening the oppositional currents in everyday environmentalism and in the emerging sustainable consumption movement. The left-wing legacy provides a theoretical and political framework that highlights both the social relations and the environmental costs embedded in the capitalist mode of production and consumption; emphasizes that individual consumption is socially structured and historically constrained; and demonstrates that this infrastructure of consumption— transportation systems, the geography of urban areas, energy technologies, communication networks, systems of health care, and so on—must be wrested from corporate control.

To these fundamental tenets of left thought, an effective green politics of consumption could fruitfully graft several familiar environmentalist themes. Greens have been warning for decades about the strains on the planet's carrying capacity caused by the consumption patterns of the affluent. Concern about social injustice as well as ecological decline is evident in their denunciations of the startling disparities in standards of living among and within nations. Green activists have also pioneered alternative definitions of the "good life," arguing that many environmental amenities can be protected only as collective goods, as part of "the commons." They have pressed challenges to uncritical acclamations of "progress" and "development," chipping away at the assumed equivalence of scientific and technological advance, economic growth, and social good.

These left and green insights can inspire both pedagogy and political action. Together they constitute a potentially powerful challenge to the consumer culture of developed capitalist societies—a challenge that could connect everyday life to global systems of production and power; that could meld calls for individual responsibility to the necessity of collective action and decision-making; that could build a powerful political vision combining environmental sustainability, social and economic justice, and a good life not available within the confines of consumer culture.

A History of Everyday Environmentalism

The green maxim that ecological devastation is disproportionately wrought by the planet's affluent consumers was evident at the first Earth Day in 1970. In the opening lines of his keynote speech, organizer Denis Hayes spoke of the disparity between the percentage of the global population living in the United States and the percentage of the world's goods and services consumed there. "We are challenging the ethics of a society that, with only 6 percent of the world's population, accounts for more than half of the world's annual consumption of raw materials," he said. Hayes went on to address the dubious conclusions often drawn, then and now, from such figures, arguing that environmental problems cannot be solved by individual lifestyle adjustments, and that such recommendations can be used to distract people from systemic and collective efforts.[8]

In those early years of the modern environmental movement, ecological scientists and activists linked "overconsumption" to the depletion of non-renewable resources and the blight of pollution. They highlighted local and regional problems—acid rain, smog-choked air, contaminated waters. By the 1990s, their core concerns about ecological limits in general, and the impacts of consumption in particular, had shifted to problems related to the waste-absorbing capacities of the planet and its various ecosystems, including the human body: climate change, ozone depletion, biodiversity loss, and the health hazards of rampant synthetic toxins.

Everyday environmentalism draws much of its vitality from this legacy of green activism, but its political and cultural history is independent, complex, and contradictory. One of the key moments in its development is the voluntary simplicity movement of the late 1960s and early 1970s, an era in which simple living was generously flavored by radical political protest and daring cultural experimentation. The voluntary simplicity movement shared with the counterculture and the women's liberation movement the notion that political commitment could and should be joined to personal fulfillment, spiritual growth, and community. Its adherents advocated a simple and "natural" lifestyle as a meaningful, healthful, and ethical alternative to the capitalist work-and-consume ethos. They also put themselves on the side of social justice and economic equity, exhorting themselves and others to "live simply so that others may simply live." *Diet for a Small Planet,* Frances Moore Lappe's 1971 bestseller, is emblematic of the period: Hunger and famine, it argues, result from social processes rather than "natural scarcity"; eating lower on the food chain will simultaneously assist the world's poor, mitigate the environmental degradation caused by meat production, and reap the health benefits of vegetarianism.

Voluntary simplicity faded together with the counterculture, but reappeared in the late 1980s in a bevy of bestselling "green lifestyle" and "simple things you can do" guidebooks. This publishing phenomenon brought everyday environmentalism to a wide audience, but also quite literally domesticated it, sequestering it in the household and supermarket. The green lifestyle guides also paved the way for a third important influence on today's everyday environmentalism: the spurious claims and relentless image mongering of the corporate environmentalists that reached a frenzy by the early 1990s.

Marketers and advertisers, product and packaging designers, and public relations experts employed by the world's largest corporations now devote huge energies and expense to greening the images of products and of corporations themselves. According to a 1995 estimate, a billion dollars a year are spent in the United States on green public relations alone.[9] Smaller entrepreneurs use green strategies to score business successes in selected market niches. Mainstream environmentalists also appeal to everyday environmentalism, liberally dispensing advice on and accessories for green living.

The 1990s resurgence of simple living, however, has been in large part a grassroots affair. Politically much more tentative than its 1970s forebear—in part because radicalism is itself so much less visible and confident than it was 25 or 30 years ago—simple living in the 1990s is also less nostalgic and less demographically homogeneous than its predecessor. Autarchic notions of going "back to the land" have dropped away, so that simple living is now as much an urban and suburban as a small-town or rural phenomenon. The new simplicity also exists in a far different economic context. It is indeed voluntary for some of its practitioners, those financially comfortable if not affluent "downshifters" who have chosen to forgo income in order to live according to different priorities and principles. Others who are attracted to the philosophy of simplicity have been prodded to adopt it by involuntary circumstances as middle-class assumptions of affluence and upward mobility have faded. Whatever initially catalyzed them, today's simplifiers worry about overwork and overspending, about "time famine" and consumer debt, and about what they see as the skewed values and dubious benefits of consumer culture—issues that in some simple living circles eclipse green concerns.[10]

The Environmental Imagination in Popular Culture

The term "environmental imagination" refers to the green ideas, images, and anxieties that so thoroughly saturate U.S. politics and popular culture in the late twentieth century. Though its presence is pervasive, the environmental imagination is somewhat elusive as an object of study. It is a sensibility rather than a clear-cut and predictable political position, or a bounded cultural phenomenon. In Gramscian terms, it is a green "common sense," a historically specific form of popular thinking that is necessarily "fragmentary, disjointed and episodic," but that is nonetheless an essential terrain of ideological and cultural struggle.[11] For all its inconsistencies, the persistence and diffusion of the environmental imagination should provoke cautious hopes for—and strategic deliberations about—the emergence of a more challenging green movement.

Like C. Wright Mills' famous sociological imagination, the environmental imagination is a "quality of mind" that can help people understand, feel, and shape their world and their place in it. Mills argued that the sociological imagination connects personal troubles to public issues, "enabl[ing] us to grasp history and biography and the relations between the two within society."[12] So too the environmental imagination. In its best and its potential forms, the environmental imagination could reveal the social arrangements that link everyday life to the fate of the Earth. It could probe a series of crucial relationships: between the natural and the social; between the local and the global; between a technoscience practiced as an arrogant elite enterprise and one that is far more tolerant and democratic; between the ecological prospects that affect all people and the disproportionate burdens that fall on the poor and the powerless.

Unlike Mills' sociological imagination, the environmental imagination is closely associated with a popular social movement. The environmental movement—itself a messy, conflictual, and unstable arena—deserves enormous credit for the creation and scope of the environmental imagination, and bears great responsibility for its shape. Among the staggering number of often divergent environmentalisms are varieties that have endowed the environmental imagination with structural and cultural riches: visions of new ways to live, work, think, and feel; innovative institutions and policy proposals; epistemological insights and and spiritual awareness; motivations for collective action toward political reform and social transformation.

Since the first Earth Day in 1970, a politically tame environmentalism has strongly asserted itself, flattening rather than enriching the environmental imagination. Mainstream environmentalists have too often emphasized the domination and exploitation of the "natural world" while barely acknowledging the domination and exploitation of human beings. They have too often politely accommodated the values and institutions of the very social system they condemn for producing global ecological crisis, proposing technocratic fixes and regulatory reform rather than basic alterations of the capitalist status quo. However, activists outside the green mainstream—grassroots antitoxics organizers, environmental justice activists, public health advocates, deep ecologists and social ecologists, ecofeminists, and many others—have also made inroads into the environmental imagination.

The environmental movement is still a key player in moving and molding the environmental imagination, but it is by no means alone. Green notions are now diffused through many filters and along myriad paths of popular culture: news stories about ecological disasters; nature documentaries, children's stories, Disney films, and action thrillers; amusement parks, shopping malls, science museums, and fashion ads; school curricula and scientific controversies; the positioning of politicians and the musings of philosophers. Corporate environmentalism, straining to soothe popular unease about the ecological costs of consumer culture, also muddies the environmental imagination—especially because so many mainstream environmentalists have chosen to cooperate with it.

Greens with a sharper political critique have worked hard to expose the inaccurate and hypocritical corporate claims of green virtue. However, in their appropriate zeal to denounce "greenwashing," most of them seem to have conceded defeat on anything that can be considered a lifestyle issue, including everyday environmentalism. In what Gramsci calls the "war of position," they have simply given up. Rather than trying to separate everyday environmentalism from its corporate seducers and nudge it in counterhegemonic directions, they decry it as co-optation—as an abject surrender to capitalist consumer culture, a total depoliticization of environmentalism, a stubborn avoidance of production and power. Of course, green marketing *is* an effort to shunt the blame for environmental problems away from corporations and capitalism, and to weaken the oppositional content of the environmental imagination. It is also, however, a tribute to the strength of the environmental movement, and a force that normalizes uneasiness about environmental conditions.

Currently, then, the environmental imagination is neither coherent nor consistent, and certainly not always radical or cleanly oppositional. Underneath and alongside environmentalist sympathies and the new green habits of daily living roil fears and desires that stymie political involvement. Impulses to confront the political and economic arrangements that devastate the environment vie with the temptation to rely on authority: to believe that *someone* must be minding the store—that the corporations, or the government, or the high-profile traditional environmental organizations will "clean up the mess." Anxieties about personal health and the well-being of the planet can feed either collective action or a fallout-shelter mentality—no contaminant will penetrate my pure and organic perimeter. Recognition of the magnitude of social change necessary to reverse environmental deterioration can breed denial or despair. Inclinations to activism often give way to feelings of isolation and helplessness.

Many of the same political and psychological muddles ensnare everyday environmentalism. More mundanely, the ordinary demands of life pose formidable obstacles to learning and following green habits. The complexity and uncertainties of ecological knowledge sometimes make it difficult to evaluate what green living entails, and no one has the time or money to research the environmental, health, and social impacts of every purchase and practice. Even when the imperatives of environmental responsibility and healthy living are clear, they are often inconvenient, sometimes extremely so. The existing infrastructure of consumption makes it difficult—often impossible—to be consistent about living ecologically; it also hides the social nature of consumption, camouflaging the corporate and governmental forces that have embedded unsustainability into the "American way of life."

Their inconsistencies and political underdevelopment notwithstanding, everyday environmentalism and the environmental imagination are firmly embedded in political and popular culture—and their insurgent edges have etched some deep marks. Environmentalists have used consumer culture as a springboard for activism, mounting boycotts, publicity campaigns, and legal battles. They have built political front lines out of everyday habits and ordinary consumer purchases: The green campaigns against ozone-depleting styrofoam and the killing of dolphins in tuna fishing are well-known examples. They have forged green images that fire the dissenting impulses of the environmental imagination: the fragile Earth floating in space, the nuclear reactor cooling tower, the garbage-laden barge turned away from port after port.

Some of these green icons point, tentatively but suspiciously, at the dominant assumptions of capitalism. The jumping dolphin on the labels of tuna cans and the toxic waste dump, for example, invoke the often painful origins and squalid fates of commodities, and disrupt unreflective assumptions of consumption without consequence. They are persistent reminders that our daily habits of shopping, consuming, and disposing are entwined in global webs of production, and inscribed in our bodies as well as in the environmental well-being of the planet.

The Politics of Consumption

Political assumptions about consumption, explicit and implicit, inevitably color assessments of everyday environmentalism. The dismissive tendencies described earlier are thus due at least in part to the traditional Left's focus on production and labor, which views consumption as a personal and household matter outside the realm of "real" politics. Though few today would endorse the gender bias that helps orient this political compass, it still finds its way into contemporary analysis. In his critique of green consumerism, for example, Timothy Luke castigates consumption as "passive"—far too easily read "feminine"—rather than "vital" like production.[13]

Around midcentury, critical theorists such as Herbert Marcuse did turn their attention to consumer culture. Recognizing its growing economic and ideological role in stabilizing capitalism, they argued (and here I sketch in exceedingly broad strokes) that people are manipulated and stupefied by advertising, that the pleasures derived from consumer culture are trivial or inauthentic, and that the co-opting power of consumer culture is a key explanation for the scarcity and ineffectiveness of radical political dissent. These critical theorists analyzed consumer culture as a bulwark of capitalism; they abhorred it for squashing individual and social creativity and for debasing cultural expression.

Though this tradition remains widely influential, considerable intellectual ferment (sometimes accompanied by considerable vitriol) now surrounds the politics of consumption on the Left. A variety of challengers, many of them associated with the academic approach known as cultural studies, have questioned the critical theorists' "manipulationist" model from a postmodern or populist standpoint. Again sketching very briefly (and in the process blurring some distinctions that fuel lengthy debates), the challengers emphasize that people are active interpreters of marketing and advertising messages. They validate many of the gratifications of consuming, but sometimes imply either that the seductions of consumer culture are completely innocent or that there can be no normative ground from which to judge them. They rearrange the map of political significance to make consumption central, but in the process often push production to the edges of their attention. They celebrate the opportunities offered by consumer culture for individual and subcultural creativity and resistance, but give scant attention to the efforts of social movements to grapple with and operate within this omnipresent cultural environment.[14]

Ironically, both sets of assumptions often seem to overlook many instances of consuming that are, in fact, work. Perhaps this is because the glitter of consumer culture and the mediagenic politics of style make the arrangements by which our material lives are reproduced seem mundane by comparison. In any case, since shopping for food and other household necessities is still associated with women, this blind spot bespeaks the need for differentiated and gendered analyses of consumption.

Though these characterizations of the competing approaches to consumer culture are far too brief to permit a thorough evaluation, they suggest the need for careful navigation through political and theoretical shoals, and for nuance and specificity in both analysis and political strategizing. They also sketch the condition of left-leaning rumination about consumer culture, against which to contrast the fragmentary but distinctive features of the green approach. In spite of the intellectual and political wrangling that surrounds consumption, the green challenge to consumer culture remains largely unappreciated. Contemporary studies of consumer culture often gesture at environmentalist ideas and projects, but rarely consider them in depth.

Environmentalists, on the other hand, have only recently begun to grapple systematically with the culture of consumption as well as its ecological consequences. Though their efforts to do so have been influenced by a variety of political, aesthetic, and religious traditions, they have remained all but isolated from recent theoretical debates about consumer culture. In part for this reason, many environmentalist discussions of consumer culture oscillate without reflection between naive optimism and unalloyed despair. Glib assumptions that consumption patterns can be transformed by moral suasion alone alternate with dire warnings that consumer culture must be immediately and if necessary coercively dismantled in order to avoid ecodisaster.

Some green commentators portray overconsumption as solely a moral or spiritual shortcoming. The widely read environmental essayist Bill McKibben, for example, pegs his analysis of consumerism to the notion of "enchantment." "[I]t's deep in our bones," he writes, "the way that religion was deep in the bones of your average fourteenth-century peasant."[15] Other environmental writers explain unsustainable consumption with medical metaphors, again obscuring the sociocultural history and political-economic forces that underlie consumer culture. Economist Paul Ekins, for example, calls consumer culture a "virus spreading across the globe."[16] Vice President Al Gore, in his environmentalist guise, characterizes consumption as an addiction—and all but prescribes a twelve-step program as the cure.[17]

Alan Durning, whose 1992 *How Much Is Enough? The Consumer Society and the Future of the Earth* was extremely influential in focusing environmentalists anew on issues of lifestyle and consumer culture, argues for dividing the world's people into three consumption classes: the consumer class (those of us who travel by private car and by air and eat meat-centered diets and processed foods), the middle class (which travels by bicycle, bus, and rail; and whose diet, based on grains, provides adequate calories and proteins, but often lacks diversity and therefore necessary nutrients), and the poor (who walk and go hungry). Though Durning's descriptions of global inequity are vivid, his notion of class is keyed only to the ecological ramifications of different patterns of daily living. It is completely divorced from considerations about justice, democracy, or political power.[18]

The Sustainable Consumption Movement

The term "sustainable consumption" joined the green lexicon in the early 1990s, catalyzed by the insistence of governments and activists from the poorer nations that "overconsumption" in the affluent North is as serious an environmental threat as "overpopulation" in the South. Sustainable consumption is now much discussed in Europe, where greens from thirty nations have launched a campaign to specify the "environmental space" that can equitably be allocated to each country.[19]

During the same period, U.S. nonprofit groups devoted to specific aspects of consumption (food, water, transportation, new economic indicators, urban or regional sustainability planning) proliferated. A "Network to Reduce Overconsumption" that was formed in 1994 boasted over 175 group and individual members a few years later. The National Religious Partnership on the Environment organized congregation-based discussion groups about what project director Rabbi Daniel Swartz calls "the marketplace [as] the main religion of America—the major idol of the twentieth century."

The formation in 1997 of a national clearinghouse on "consumption, quality of life, and the environment" was a significant development for U.S. sustainable consumption efforts. Blending appeals for sustainable consumption and simple living, the Center for a New American Dream (CNAD) has set out to demonstrate that consumer culture threatens not only ecological but also personal well-being. The "new American dream" that it promotes is one unlinked from conventional assumptions about the meaning of economic growth, the virtues of ever higher levels of personal income for the "consuming classes," and the necessity for ever longer hours of work.[20]

This organizational surge suggests the emergence of a new political force, a movement for sustainable consumption. The next several years will be crucial ones for the U.S. sustainable consumption movement, in all likelihood determining the scope of its constituency and the tone of its politics. How will sustainable consumption activists forge the connections between personal and political change? How will they confront the efforts of transnational corporations to ensure sustainability on *their* terms? Will they support green taxes that are regressive? Foster resignation to decaying living standards in the name of environmental necessity? To what extent will they emphasize the quality of life of the middle classes over attention to the often desperate lives of the poor?

Some of the signs are discouraging. Sustainable consumption activism often fails to embrace grassroots efforts that could easily be considered part of its purview. The community food security movement, antisweatshop organizing in the garment industry, opposition to suburban sprawl and to cuts in public transit, and environmental health projects, for example, aren't widely seen as efforts toward sustainable consumption. Nor do these grassroots movements necessarily identify themselves that way. It is quite feasible, however—and certainly politically important—that meaningful links could be formed between sustainable consumption activities that agitate for redefining the "good life" and those that focus on the everyday conditions of the less affluent, many of whom should be helped to afford more rather than encouraged to consume less.

Some of the key figures in the emerging sustainable consumption movement are well aware that attention to equity and justice must be incorporated into their strategies. CNAD director Betsy Taylor, for example, has spoken of "the centrality of justice to the issues of consumerism and sustainability" and argued that sustainable consumption has the potential to serve as "a bridge between those who have a lot and those who have not enough."[21] Unfortunately, other individuals and groups working toward sustainable consumption seem untroubled by the class bias of their organizing. In a recent online conversation sponsored by CNAD, several participants explicitly argued that sustainable consumption efforts focus mainly on "middle to upper-income suburban families across the United States."[22]

If a U.S. sustainable consumption movement continues to develop, political debates such as these will have to be joined and strategic decisions reached. Sustainable consumption activists will have to decide whether to pursue a dead-end path toward a tepid lifestyle environmentalism, or begin a far more challenging journey. Ignoring the plights of the working and nonworking poor would make the sustainable consumption movement irrelevant at best, and might well aggravate class and racial tensions within U.S. environmentalism.

A radical and effective sustainable consumption movement would need to offer social visions and political programs that simultaneously protect the planet, redefine and improve quality of life, promote social and economic equity, and challenge corporate control and capitalist hegemony. Instead of addressing themselves exclusively to the middle classes, advocates of sustainable consumption and green living should ally themselves with those who have been shut out of the old American dream, and whose active support is needed to forge a new one.

Sustainable consumption activists will also soon face rancorous debates about the "American way of life" that are sure to accompany efforts to reduce production of greenhouse gases. A climate change treaty can succeed only if the wealthier nations agree to deeper cuts in carbon emissions than are required of the poorer ones. Opponents of such differential reductions—along with climate change skeptics who simply refuse to see a problem—will issue ominous warnings about "freezing in the dark" and echo President Bush's myopic declaration at the 1992 Earth Summit that "the American way of life is not up for negotiation." Sustainable consumption activists are well positioned to rebut these tactics of belligerent denial, and to insist on the urgency of environmental restoration *and* social justice, while offering compelling alternative visions in which the good life is, in part, a life of political engagement.

The green challenge to consumer culture is a work in progress. Everyday environmentalism and the environmental imagination in which it is embedded harbor fertile seeds of opposition to consumer culture. They are precious resources for activists working to build a sustainable consumption movement strong enough and smart enough to tackle the enormous changes that our environmental and political predicaments demand. In the context of a robust and radical movement, they could become nodes of collective opposition to the culture and structure of capitalism. In the absence of compelling visions and strategies for fundamental social change, these precious resources will be squandered.

Notes

1. On simplicity circles, see Cecile Andrews' *The Circle of Simplicity: Return to the Good Life* (New York: HarperCollins Publishers, 1997). Eco-teams, hundreds of which have been organized by a group called Global Action Plan, are small groups of friends, neighbors, or co-workers who agree to work together to green their households or workplaces. "Building the Voluntary Simplicity Movement" was part of the title of a conference called in September 1998 by the group "Seeds of Simplicity."

2. The Harwood Group, *Yearning for Balance: Views of Americans on Consumption, Materialism, and the Environment* (Takoma Park, Md: Merck Family Fund, July 1995).

3. Juliet B. Schor, *The Overworked American: The Unexpected Decline of Leisure* (New York: Basic Books, 1992).

4. Tom Athanasiou, *Divided Planet: The Ecology of Rich and Poor* (New York: Little, Brown, 1996).

5. Brian Tokar, *Earth for Sale: Reclaiming Ecology in the Age of Corporate Greenwash* (Boston: South End Press, 1997, p. xiii.

6. Timothy Luke, "Green Consumerism: Ecology and the Ruse of Recycling" in *In the Nature of Things: Language, Politics and the Environment,* Jane Bennett and William Chaloupka, eds. (Minneapolis: Univ. of Minnesota Press, 1993, pp. 158–59).

7. Eugene Coyle, Topic 98, online Peacenet conference *env.consumption,* 1996. Athanasiou, Coyle, and I are members of a San Francisco Bay Area political ecology study group, and have discussed our differences about environmentalism and consumer culture at length and fruitfully. I thank them and others in Lorax for their articulate and intelligent feedback.

8. Denis Hayes, keynote speech in *Earth Day—The Beginning: A Guide for Survival,* Environmental Action staff, eds. (New York: Bantam Books, 1970, p. xiii–xv).

9. John Stauber and Sheldon Rampton, *Toxic Sludge Is Good for You: Lies, Damn Lies, and the Public Relations Industry* (Monroe, Me: Common Courage Press, 1995).

10. See Juliet B. Schor, *The Overspent American: Upscaling, Downshifting, and the New Consumer* (New York: Basic Books, 1998) and *The Overworked American: The Unexpected Decline of Leisure* (New York: Basic Books, 1992); also Arlie Hochschild, *The Time Bind: When Work Becomes Home and Home Becomes Work* (New York: Metropolitan Books, 1997).

11. Cited in Stuart Hall, "The Problem of Ideology: Marxism Without Guarantees," in *Stuart Hall: Critical Dialogues in Cultural Studies,* David Morley and Kuan-Hsing Chen, eds. (New York: Routledge, 1996 [1983]): p. 43.

12. C. Wright Mills, *The Sociological Imagination* (New York: Oxford Univ. Press, 1959), p. 6).

13. Luke, "Green Consumerism," p. 158.

14. The recent literature on consumption and consumer culture is vast. For a few representative texts, see Frank Mort, "The Politics of Consumption," in *New Times: The Changing Face of Politics in the 1990s,* Stuart Hall and Martin Jacques, eds. (New York: Verso, 1990), pp. 160–172; Don Slater, "Consumer Culture and the Politics of Need," in *Buy this Book: Studies in Advertising and Consumption,* Mica Nava, Andrew Blake, Iain MacRury, and Barry Richards, eds. (New York: Routledge, 1997), pp. 51–63; Martyn J. Lee, *Consumer Culture Reborn: The Cultural Politics of Consumption* (New York: Routledge, 1993); Daniel Miller (ed.), *Acknowledging Consumption: A Review of New Studies* (New York: Routledge, 1995).

15. Bill McKibben, *Maybe One: A Personal and Environmental Argument for Single-Child Families* (New York: Simon & Schuster, 1998, pp. 114, 116).

16. Paul Ekins, "The Sustainable Consumer Society: A Contradiction in Terms?" *International Environmental Affairs* 3(4) (Fall 1991), pp. 243–258.

17. Al Gore, *Earth in the Balance: Ecology and the Human Spirit* (New York: Plume, 1992).

18. Alan Durning, *How Much Is Enough? The Consumer Society and the Future of the Earth* (New York: W. W. Norton, 1992).

19. Two recent books describe European approaches to sustainable consumption and ecological space: Michael Carley and Philippe Spapens, *Shaping the World: Sustainable Living and Global Equity in the 21st Century* (London: Earthscan, 1998) and Wolfgang Sachs, Reinhard Loske, Manfred Linz et al., *Greening the North: A Post-Industrial Blueprint for Ecology and Equity* (New York: Zed Books, 1998).

20. See the CNAD Web site, www.newdream.org.

21. Betsy Taylor, author's interview (November 27, 1995).

22. For an exchange on this point between Michael Wood-Lewis and myself, see the archives of this online discussion on the CNAD web site: www.newdream.org/conversation-arc/msg00100.html, www.newdream.org/conversation-arc/msg00108.html, and www.newdream.org/conversation-arc/msg00111.html.

13

Rethinking Technoscience in Risk Society: Toxicity as Textuality

Timothy W. Luke

This chapter explores some of the ways in which the scientific facts of toxicity and broad risks to human health are turned into contestable texts for scientific experts and the lay public by environmental toxicology, ecological risk assessment, and public health science. Before anyone can organize an effective strategy for coping with increasing levels of toxicity and their effects on human and nonhuman life, toxics must be identified, measured, and interpreted by those who use them as well as by those who are affected by them.

The discourse on their dangers, which arises out of public and private efforts to understand and manage them, often conceals some threats even as it reveals others. As Steingraber and O'Brien argue in this volume, toxics, along with the larger understanding of their effects, are socially produced, and the processes of evaluating their costs and benefits frequently use scientific evidence in disinformative ways. Because a more sophisticated science of analyzing toxicity is not yet at hand, our understanding of toxics often is much more explicitly textual than it is technical. A text is whatever can be read or reinterpreted beyond some recognized conventional meaning, and textuality marks anything that evokes, or is seen as capable of generating, many successive rereadings and interpretations. Toxic substances are constantly subjected to such moments of interpretation, but few of these readings go unquestioned. Consequently, a convoluted chain of textuality rattles through every question raised about measures of toxicity or every challenge to the meaning of such measurements.

To see toxicity as textuality is to admit to the contested, unknown, and indeterminate qualities of toxic effects, which we are not able to identify effectively with any real foresight, to measure with complete accuracy, or to interpret reliably with total validity. Assessment of environmentally toxic substances by technoscience communities as well as ordinary lay persons soon boils down, just as Levenstein and Wooding suggest in this volume, to a game of open-ended textual analysis rather than a process of obtaining closed technical certainties. Viewing toxicity as textuality, in turn, could disclose much larger social forces at work in the toxic waste problem. This would allow us to better understand how some interests seek to distract with corporate science, downplay dangers with high technology, and disinform with official ideology, while other groups contest scientific practices, criticize quick technofixes, and counter widespread misunderstandings.

Treating toxicity as textuality also highlights the equally problematic dangers that unfold within the professional-technical practices of environmental toxicology, risk assessment, and public health, which are being mobilized as scientific resources for the conduct of civic life. All too often the type and quantity of evidence needed to assess environmental risk is treated as purely scientific data secure on allegedly pure technical grounds, when, in fact, determining the qualities of the toxicity often is a *social act of political interpretation* as much as, if not more than, it is a mechanical procedure for assaying the acceptability of a risk. Toxicity is thus always inescapably textual: it never exists simply as such, and does not reveal all of its negative effects directly.

Instead, it is a communal construct—fearing it, seeing it, typing it, measuring it, judging it—all involve many complex, multilayered acts of cultural, political, and social interpretation. Approaching toxicity as something that can be determined through mechanistic operations of scientific measurement and analysis makes a complex process look simple, replaces multiplicity with simplicity, and derives results from questionable evidence in a fashion that is both inept and unjust.

This chapter attempts to understand toxic wastes as a fixed feature rather than a manageable passing mishap in modern economies and societies. Sometimes by accident and sometimes by design, toxic wastes have become part of a long line of industrial by-products that emerge alongside ordinary industrial production. Whether in the factory or the field, toxic contaminants are permanent new qualities of the Earth's modernized ecologies.[1] By reevaluating the textualities of toxicity, this analysis argues that toxic wastes constitute a key part of the background conditions for all human and nonhuman life. The technoscientific practices that intentionally produce commodities also unintentionally create cancer, and the essential goals of corporate economic growth bring with them ungovernable increases in toxic by-products that create death, disease, and disability as part of nontoxic products.

As Steingraber asserts, the social production of cancer cannot be separated from the social production of commodities. Without any clear strategy for mounting a more powerful collective response, liberal society mostly puts the responsibility for assessing the risks and managing the threats of toxics onto the individual, turning the containment of unreasonable ecological costs in the marketplace into a personal health benefit created by private responses at home.

Toxicities and Textualities

Not all chemicals are, in and of themselves, intrinsically contaminants or toxic as such. Instead, as Lappé maintains, "a chemical becomes a toxic contaminant when it is not naturally found in a specific locale; it is potentially harmful; and when its level exceeds a permissible threshold, usually set at or below a concentration where exposure can be expected to produce 'no observable effect' on health. For chemicals that are carcinogens, this level is often set at the lowest concentration feasible or, preferably, zero."[2]

All of these technical characteristics often seem scientifically self-evident, but in fact are not obvious. Gauging toxicity depends on many different elements that must be scanned and then decoded like a text. What is the substance? How is it naturally found? Where is some specific locale in which it is discovered? How is its toxic potentiality judged? What is a permissible level of contamination? When is an observable toxic effect sensed? This interpretive problem is textual and contextual rather than self-evident and clear.[3] A toxic substance is produced, but in circumstances that amplify or mitigate its effects. Still, as it becomes enmeshed in codes of scientific analysis and subcodes of public interpretation, an understanding of toxics, like all textuality, cannot be easily derived from the codes of analysis used in their interpretations. The context, the interpreter, and the audience usually interact in ways that produce disagreement over what is a contaminant, which material is unnatural, what is a permissible threshold, what amount of exposure is dangerous, and what is an observable effect.

"Toxic wastes," as Smith suggests, "are a by-product of energy development, agriculture, and most industrial activity," which now "are found throughout the environment, in our air, water, and soil."[4] Each of the Earth's industrial economies produces these noxious outputs as inherent by-products of ordinary everyday life. Yet the U.S. Office of Technology Assessment believes that "there are major uncertainties on how much hazardous waste has been generated, the types and capacities of existing waste management facilities, the number of uncontrolled waste sites and their hazard levels, and on the health and environmental effects of hazardous waste releases."[5]

These kinds of observations highlight the ubiquity, opacity, and complexity of interpreting the toxic impact of hazardous waste on human and nonhuman life. Like the air, water, and wildlife, agricultural and industrial wastes are to be found everywhere in the environment, making them a new fundamental characteristic of the Earth's ecology as it is transformed by modern agricultural, industrial, and technological development.[6] The many mechanisms that place chemicals outside specific locales, boost their concentrations beyond permissible thresholds, produce exposures so intensive that they threaten health, and disperse effects indiscriminately across space and time are human creations. Some by-products are intended and their effects are understood, but even their producers cannot completely anticipate or comprehend all of the effects of these substances.

Toxicological science is always trapped between technology and textuality as it conducts its environmental studies. On the one hand, toxic substances pose a very real threat to human well-being, but they also give rise to their own special problems.

They are invisible, typically undetectable intruders that can harm us in ways we might not discover for years. We are ignorant for the most part of the mechanisms by which they are transmitted and by which they harm us, which makes their causal path difficult to trace. Many have the potential for catastrophic consequences at least to the affected individuals, yet typically they are associated with modest benefits. The catastrophic injury, however, frequently has a low probability of occurring.[7]

The technical measurements of toxicity are always ambiguous, and the human response to them presents very special problems because of the complex ways chemicals can interact in human bodies, the natural environment, and nonhuman life.

Because of the invisible presence, long latency periods, obscure mechanisms, and untraceability to responsible agents, in assessing the risks from carcinogens we are forced to spell out the "faithful signals" of the fragmentary truth about carcinogens even when "they seem obscure to us." And we must try to detect the signals in a timely fashion before the threats materialize into harm. However, several of the scientific fields that might enable us to discover and accurately assess such risks are in their infancy. Furthermore, the demanding standards of evidence and requirements for certainty commonly considered part of research science frequently frustrate our efforts to identify, regulate, and control toxic substances. A more serious problem is that such procedures are much too slow to evaluate adequately all of the substances in commerce or that enter commerce daily.[8]

Environmental toxicology on a planetary scale, given these textual complexities, the difficulty of interpreting results, and the ambiguous evidence, is not a decisively accurate enterprise. Even though many of its proponents act as if its findings provide enough understanding to guide individual and group decision-making, its techniques are always eclipsed by textualities.

Steingraber also imagines toxicological analysis as a textual process of sorts. "Our bodies," like ancient trees that can be cored for dendrochronological assays, "are living scrolls of sorts. What is written here—inside the fibers of our cells and chromosomes—is a record of our exposure to environmental contaminants. Like the rings of trees, our tissues are historical documents that can be read by those who know how to decipher the code."[9] Burdened with trace amounts of innumerable substances, this somatic scroll, or *body burden*, refers to the sum total of these exposures and includes all routes of entry (inhalation, ingestion, and absorption through the skin) and all sources and places of exposure (food, air, water, workplace, home, and so forth). The somatic scroll is a key text for toxicologists to decipher.[10] The nuances of such readings are frequently ignored, but they are real, even when they are disguised in technical routines. Detecting traces of exposure, tracing their effects all the way back to a source, and identifying strategies to halt this contamination are practices that ultimately are partly science and partly art.

The interplay of textuality and technique in environmental toxicology soon acquires confounding complexity.[11] Lappé concedes as much when he wonders if science can make "accurate predictions about the composite impacts of new chemicals and electromagnetic radiation on all living systems." When even a few factors are taken into account, predictions become quite hazy. For example,

in exposure to ultraviolet light that is expected as a result of sunlight passing through our chemically fragmented ozone layer. . . . Where will these powerful rays first impinge on living things? How will their effects be felt on ecosystems that depend on sunlight for their energy? How will humans, especially those whose skins are highly permeable to light, react? And how will ultraviolet light interact with the thousands of chemical contaminants that permeate the environment?[12]

Our survival may well depend upon understanding these sorts of reactions, but the sciences of biology, epidemiology, pathology, oncology, and toxicology cannot provide any definitive guidance. They do not have complete command of the data, techniques, and resources needed first to ask and then to answer all of the questions. Those questions that can be asked and answered often look at tests on only one or two compounds, mostly in animal or plant models. Bigger questions about the effects of many compounds working together in humans often cannot be modeled effectively. For most chemicals of concern, as Lappé himself admits, the science of risk assessment is hamstrung by unknown interactive effects and by animal tests that are only indirectly predictive.[13]

For almost any substance in the environment, toxicity becomes an open-ended textual question rather than a purely closed technical one. A toxin loose in a particular biota or biosystem produces a web of effects whose many threads cannot all be accounted for or fully understood, even if they can be found. The disposition, or impact of a toxic substance on a life form can only be measured in dosages and exposures that blend bits of incomplete evidence to produce synthetic interpretation of results.

Trying to make an exact science of environmental toxicology quite often is nothing more than a gross affectation predicated upon mostly unfounded extrapolations from controlled studies of hazardous materials in laboratory settings. Given this, science often leaves us virtually helpless and adrift on the seas of disinformation that O'Brien fears. In turn, the conventional response to toxic contaminants, as Buchholz[14] illustrates, reads their textual signs either as instances of new regulatory supervision by the state or opportunities for new commodification. The poisoning and polluting systems of industrial by-production are seen as new opportunities for "the greening of business" as corporations mobilize their profit-seeking principles of environmental management.

Technoscience and Toxic Wastes

The development of new technoscience disciplines, such as environmental toxicology, risk assessment, or public health, mark a shift from unreflective industrial development to a more reflective one of risk management amid the uncertainties of a modernized ecology. As Beck suggests, the risk society is here, and a general public health strategy guided by environmental toxicologists should confirm for any citizen how "the social production of *wealth* is systematically accompanied by the social production of *risk*" and how "the problems and conflicts relating to distribution in a society of scarcity overlap with the problems and conflicts that arise from the production, definition, and distribution of techno-scientifically produced risks."[15]

With all of its practical involvement in public health administration and environmental science, toxicology tacitly indicates how the economic imperatives behind technological development are now "being eclipsed by questions of the political and economic 'management' of the risks of actually or potentially unlisted technologies—discovering, administering, acknowledging, avoiding or concealing such hazards with respect to specially defined horizons of relevance."[16] With this recognition, the toxicity of many substances—industrial by-products, agricultural chemicals, construction materials, artificial foodstuffs, nuclear waste, automotive fuels, food packaging, synthetic pharmaceuticals, to name only a few—becomes a contested text, brimming with actual and/or potential hazards awaiting further analysis and interpretation by experts and lay persons alike.

Lappé observes, "we are in the midst of the chemical revolution. It is a given that the chemical industry and its allied field of pharmaceutical and pesticide manufacture represent dominant forces that are shaping our world. . . . that chemicals insinuate themselves into our lives."[17] Without saying so directly, Lappé confirms how thoroughly revolutionary the ensembles of chemical science, chemical industrialists, and chemical manufactures are to the extent they can refashion many human-ecology relations without much effective political regulation. Under these revolutionary conditions, industrial production and its by-products now constitute humanity's most immediate environment as we evolve with unstable technoscience artifacts set beneath, within, and above each nation-state still being organized around our political acts.

Beck's analysis of "subpolitics" in the risk society underpins this analysis. More specifically, the restructuring of human lives around the ill-effects of chemical, industrial, and nuclear wastes proves how

the potential for structuring society migrates from the political system into the sub-political system of scientific, technological and economic modernization. A precarious reversal occurs. *The political becomes non-political and the non-political political.* . . . The promotion and protection of "scientific progress" and of "the freedom of science" become the greasy pole on which the primary responsibility for political arrangements slips from the democratic system into the context of economic and techno-scientific non-politics, which is not democratically legitimated. A *revolution under the cloak of normality* occurs, which escapes from possibilities of intervention, but must all the same be justified and enforced against a public that is becoming critical.[18]

The toxic waste problem is but one facet, albeit a quite harmful one, of how we live uneasily with technoscientific modernization. Even though we do not want them, do not vote to get them, and do not wish to experience their side effects, the toxic by-products of chemical transformation insinuate themselves into our lives, because we accept them with the purchase of every bug bomb, paint thinner, synthetic antibiotic, or artificial sweetener brought to us by industrial production.

Consumer markets accept these innovations because such improvements are believed to bring the good life, albeit at times with a few unwanted but allegedly controllable noxious by-products. In fact, however, technoscientific artifacts increasingly undercut the workings of conventional political life. Beck argues that the unintended effects of such choices guarantee that

political institutions become the administrators of a development they neither have planned for nor are able to structure, but must nonetheless justify. . . . What we *do not* see and *do not* want is changing the world more and more obviously and threateningly.[19]

Environmental toxicology makes the same point about the modern chemicalization of everyday life. Under the cover of normality within industrial production, what we do not see and do not want from industrial by-production is obviously changing our world quite thoroughly. The toxicological studies conducted by environmental public health authorities try to overcome the negative effects of anonymous decisions that change society by quantifying the incidence, level, and severity of the risks produced by technical modernization in the new narratives of "public advisory" reports. However, this is not enough.

In these conditions, a conventional politics of voting, polling, or organizing often fails. The chemical revolution with all of its toxic by-products is highly technified economic action that always "remains shielded from the demands of democratic legitimation by . . . a third entity: economically guided action in pursuit of interests."[20] Lappé worries that the toxic threats to public health are "not simply going to disappear into the night," even though everyone has begun "to realize that how chemicals are made, packaged, and discarded radically affects our lives."[21] Because of this and all of the other technical revolutions that will continue, "nothing less than a science of planetary toxicology is needed to measure the results of chemical insults to the thin web of life on Earth. . . . The window into this world of real and imagined peril is the science of toxicology."[22]

Even so, the inhabitants of modern states, who must live under this toxicological cloud, have yet to admit how "the structuring of the future takes place indirectly and unrecognizably in research laboratories and executive suites, not in parliament or in political parties."[23] The small bits of information that the public can gather become part of the textuality of toxicity, and this is all that they have for judging the human costs of chemical revolution, environmental transformation, and technological innovation. Technoscience may make better living through chemistry possible, but this new benefit also entails the planned and unintended destruction of many nonhuman and some human lives.

When put into practice, most environmental risk analysis unfortunately serves as an applied science of mortality management. To coexist with the techniques of wealth production, everyone consents to coevolve with the tools and techniques that generate hazardous by-products as part of their useful products. So that many might live more fully with the manufactured goods and services that insinuate their way into our lives, a few must die and/or live less fully. Just as the state often must conscript its members to wage war and die for its survival, the market accepts this arrangement for random decimation of its members in order for it to continue developing. To enjoy the production of wealth by advanced technologies, everyone must endure the systemic by-production of greater risks, recognizing that for every A, B, or C benefit of this chemical or material, X people per 10,000, Y people per 100,000, or Z people per 1,000,000 will be harmed by ill health, genetic mutation, and/or death.

Accepting these effects of advanced technologies does not move modern society very far past the bargains struck by crude rituals of human sacrifice. In modern society, everyone tacitly consents to the crippling and painful execution of many of their fellow consumers every time they spray herbicide on lawns, fill their gas tanks with high-test, buy pressure-treated lumber, and purchase plastic housewares. Statistics can forecast in general how many people will be struck by this anonymous violence, but no estimation technique or modeling trick can indicate which individuals will be taken by this brutal regimen of inexorable random decimation.

Dealing with socially produced risks in this fashion normalizes the creation of such general effects within any particular economy and society. Because the technoscience that creates and contains such by-products will not change, everyone must, on the one hand, resign themselves to the fact that such dangerous by-products are a fixed feature of the products delivered to them by industrial development. On the other hand, recognizing that science might deliver fairly reliable probabilistic statements about the types of health effects, their relative severity, and the population that will be affected simultaneously naturalizes risk (turns it into an unavoidable background condition), socializes it (reduces it to a collective cost born by all), and personalizes it (transforms it into a matter of life-style choice).

The regulation of toxic substances, then, is another manifestion of technological normalization. Acceptable levels of risk are markers that indicate the range of normality and abnormality. Toxic wastes, industrial pollutants, and biological hazards are normalized by defining their abnormalities. The portrayal of risk as a function of lifestyle and personal initiative makes toxic substances a surmountable obstacle that rational individual choices can overcome. "So we see," as Canguilhem suggests, "how a technological norm gradually reflects an idea of society and its hierarchy of values, how a decision to normalize assumes the representation of a possible whole of correlative, complementary or compensatory decisions."[24] Risk analysis creates the health advisories, and citizens thereby become the advised, struggling to determine the path of maximum likely survival from a stream of health news, food scares, toxic alerts, and hazard warnings. Nonetheless, as the U.S. Office of Technology Assessment affirms, this strategy requires an inventive interpretation of all this information by everyone, because there are no indisputable advisories based on certain, reliable knowledge about the impact of any toxic waste problem.

Managing Toxicity: The Freedom to Choose

While toxic wastes can be found everywhere, they are most easily discovered in a few places, particularly those inhabited by the poor, racial minorities, or powerless ethnic groups, who are all neglected by the larger majority in society. As Bullard asserts, these people often are considered "throw-away communities," and their lands are used for garbage dumps, transfer stations, incinerators, and other waste disposal facilities.[25] The environmental justice community opposes this sort of environmental discrimination by insisting upon a new concern for social equity and distributive impacts[26] in evaluating the negative effects of industrial by-products. However, it cannot succeed solely by shifting the focus of mainstream environmentalism, or "protecting the environment from humans," to a simple form of environmental justice, or "protecting humans from the environment."[27]

Because we have not protected nature from humans, it is now different in many respects—it has become a "modernized ecology." Therefore we must protect all humans and the not yet fully contaminated environment from those modernized ecologies already suffering the effects of extensive industrial by-products. To attain environmental justice, just environmentalism, as we have defined it thus far, is no longer enough. Instead, the regimes of governance that permit these inequities to develop must be reassessed and then reconstructed. In defining the qualities of environmental toxicity and nontoxicity, experts are given considerable authority. As Canguilhem argues, "a normative class had won the power to identify—a beautiful example of ideological illusion—the function of social norms, whose content it determined, with the use that class made of them."[28] Environmental regulations, toxic waste controls, and biohazard guidelines, as experts construe them, only push ecological abnormalities into the realm of other more stable juridical norms, such as economy, efficiency, or equality. There government, as Foucault argues, tries to operate by organizing "the right disposition of things, arranged so as to lead to a convenient end."[29]

On one plane, technoscientific expertise might locate a normal measure of government in the right networks as they enact a disposition of things and men. These networks might be rearranged constantly to lead to more convenient ends, but on another plane, experts also must articulate new understandings of abnormality when markets and/or states create wrong outcomes in the dispositions of convenience between things and men. Much of the friction, risk, and irrationality of toxic waste flows from seemingly impenetrable and inconvenient ends in the relationships between things and people. Often the right disposition of people and things in one set of technological ties creates a wrong disposition between one group of people and another or one group of things and other things in many different markets. This disposition results in polluting, toxic, biohazardous sets of relations.

The rightness or wrongness of these dispositions today remains clouded by the rhetoric of personal freedom. Liberal philosophies of agency and society often have purposely intertwined themselves with toxicity, waste, and risk in an unproductive fashion, all in the name of more choice and less regulation. The basic premises of liberal society suggest that all individuals are rational agents who, once armed with adequate information or understanding of some problem, can decide upon a suitable strategy for coping by themselves.

Risk assessment and management constantly promote such liberal responses to environmental health risks, but the indeterminate nature of toxicity is rarely made clear. The characteristics of most toxic threats are such that there neither will be sufficient information nor can there be any adequate basis for complete understanding for anyone in the available texts of toxicity. Liberal advocates of individual decision-making will retort that nothing can be guaranteed with absolute certainty. This might be true, but it also turns every collective response to modern technological disservices, like the growing toxic threat, into a crap shoot.

Because those who suffer can exert significant pressure within the political process, many state institutions have had to construct systems of risk assessment and management, to respond to this political pressure. While the reality of environmental inequality leads many to believe that many of these policy mechanisms are purely symbolic, they are not.

Cranor asserts that risk assessment works as an ad hoc alliance of vigilant regulators and informed citizens reading the textuality of toxicity together. "*Risk assessment* aims at providing accurate information about risks to human beings so that agencies in fulfillment of their statutory mandates can regulate exposure to potentially carcinogenic substances," while "*risk management* is concerned with managing the risks in accordance with statutory requirements and other economic, political, and normative considerations."[30] At this time, there is neither perfect information about actual levels of harmful effects nor reliable estimates of probable harmful outcomes. Instead, questions about dosage amounts and exposures, animal studies, and contaminant sources all generate only more questions, which makes it quite difficult to provide anything more than risk and uncertainty assessments that are "the third-best solution to a harm assessment."[31] Therefore, any regulatory decision by the state involves identifying a hazard, assessing its toxicity, gauging its probable harm, mobilizing control options, and calculating ultimate risks. These bureaucratic maneuvers are complicated because "different kinds and amounts of *evidence* may be needed before one asserts for purposes of *understanding* the definitive existence of causal claims versus *deciding* for public health purposes what to do."[32] Thus, risk management often is little more than a promise to keep writing more risk advisories.

When the evidence seems more compelling, as with dioxin or plutonium, the state tends to be very restrictive in its controls. For most toxic by-products in industrial markets, the state allows, first, for the "people's good judgment and the forces of the economic market to guide exposure," and, second, if injured, people "can take care of themselves through the tort law when injured by seeking compensation from the wrongdoer for their injuries."[33] When the risks are so compelling that some probable harm of considerable magnitude will occur, the state might try to eliminate some toxic materials almost completely. However, dioxins and plutonium still circulate in the world economy. Consequently, the management of highly socialized costs is privatized and legalized in operational codes that implicitly see toxicity as textuality. By requiring greater citizen awareness, consumer information, and customer knowledge, the market notifies everyone of their risks in arcane texts about toxicity. What cannot be known or controlled is then dealt with through tort actions, administrative laws, or regulatory interventions.

Touting the merits of liberal freedom sadly is not an uncommon response of both states and companies in contemporary capitalist systems, but it also is proving to be the most efficient mode of externalizing the costs of the toxic by-products of an industrialized economy.

To conclude, textualizing the understandings of toxicity leaves every citizen and consumer more aware of their roles as lay readers struggling either to decipher useful information from public toxic advisories or to decode the final meaning in scientific hieroglyphics about environmental hazards. A cautious reading leads to one outcome, a looser one leads off in another direction, a non-reading results in disaster. These interpretative moments, however, often shift responsibility for the first moves of detection and remediation from toxic-producing firms to toxic-consuming publics, and hides the banal plot of profit-seeking action in the liability-evading sciences of ecological disaster. Plainly, a more sophisticated science of reading toxicity is not yet at hand, and everyone knows all too well that their own complicated ties to toxics in the marketplace cannot be divorced from every catastrophic toxic event.

Notes

1. Timothy W. Luke, *Ecocritique: Contesting the Politics of Nature, Economy, and Culture* (Minneapolis, Univ. of Minnesota Press, 1997).

2. Marc Lappé, *Chemical Deception: The Toxic Threat to Health and the Environment* (San Francisco: Sierra Club Books, 1991, p. 277).

3. A text, as Eco maintains, is whatever can be interpreted as a network of different messages depending on different codes and working at different levels of signification. Umberto Eco, *The Role of the Reader: Explorations in the Semiotics of Texts* (Bloomington: Indiana Univ. Press, 1984, p. 5).

4. Zachary A. Smith, *The Environmental Policy Paradox*, 2d ed. (Englewood Cliffs, N.J.: Prentice-Hall, 1995, p. 170).

5. U.S. Office of Technology Assessment, *Technologies and Management Strategies for Hazardous Waste Control* (Washington, D.C.: U.S. Government Printing Office, 1983, p. 13).

6. National Academy of Sciences, National Academy of Engineering, *Technology and Environment* (Washington, D.C.: National Research Council, 1989).

7. Carl F. Cranor, *Regulating Toxic Substances: A Philosophy of Science and the Law* (Oxford: Oxford Univ. Press, 1993, p. 3).

8. Ibid., p. 3.

9. Sandra Steingraber, *Living Downstream: An Ecologist Looks at Cancer and the Environment* (Reading, Mass.: Addison-Wesley, 1997, p. 236).

10. Ibid.

11. Lappé, *Chemical Deception*, p. 3.

12. Lappé, *Chemical Deception*, p. 3.

13. Lappé, *Chemical Deception*, p. 7.

14. Rogene A. Buchholz, *Principles of Environmental Management: The Greening of Business* (Englewood Cliffs, N.J.: Prentice Hall, 1993, p. 286–305).

15. Ulrich Beck, *The Risk Society: Towards a New Modernity* (London: Sage Publications, 1992, p. 19).

16. Ibid., pp. 19–20.

17. Lappé, *Chemical Deception*, p. 1.

18. Beck, *Risk Society*, p. 186.

19. Ibid., p. 187.

20. Ibid., p. 222.

21. Lappé, *Chemical Deception*, p. 1.

22. Ibid., p. 2.

23. Beck, *Risk Society*, p. 223.

24. Georges Canguilhem, *The Normal and the Pathological* (New York: Zone Books, 1991, p. 246).

25. Robert D. Bullard, *Dumping in Dixie: Race, Class, and Environmental Quality,* 2d ed. (Boulder, Colo.: Westview Press, 1994, p. xv).

26. Ibid., p. 3.

27. Ibid., p. 139.

28. Canguilhem, *Normal and Pathological,* p. 246.

29. Michel Foucault, *The Foucault Effect: Studies in Governmentality.* Graham Burchell, Colin Gordon, and Peter Miller, eds. (Chicago: Univ. of Chicago Press, 1991).

30. Cranor, *Regulating Toxic Substances,* p. 13.

31. Ibid., p. 14.

32. Ibid., p. 46.

33. Ibid., p. 49–50, 51.

III

Notes from the Field: Community Struggles

14

Silencing the Voice of the People: How Mining Companies Subvert Local Opposition

Al Gedicks

The political struggles over mining projects in or near Native American communities are about survival—the protection of human health, the culture of a people, and the preservation of the ecosystem. Equally important, as detailed here, they are struggles about democracy, as large corporations seek to exercise their power without effective public participation. In 1993 the provincial government of British Columbia decided to safeguard a vast northern wilderness from the ravages of mining by designating a 2.5-million-acre watershed of the Tatshenshini and Alsek rivers a provincial park. The original proposal called for shipping the ore 150 miles by truck or slurry pipeline to the deepwater port at Haines, Alaska. Among those opposing the project were the Chilkat tribe at Klukwan village near Haines. Tribal leaders feared that any ore spills would drain into the Chilkat river and threaten their main source of food—the salmon fishery.[1] This decision effectively halted plans to build the hemisphere's largest open-pit copper and gold mine, the $430 million Windy Craggy project.

In 1997 President Clinton announced the cancellation of a huge $650 million gold mine near the border of Yellowstone National Park in Montana. Grassroots environmental groups said that acid mine drainage was inevitable because of the highly acidic ore. Moreover, the permanent storage of toxic mine waste at the proposed New World Mine would forever threaten fish, wildlife, and water quality in the area, as well as human health. The cancellation was the culmination of a bitterly contested five-year battle to halt the project led by the Greater Yellowstone Coalition and the Beartooth Alliance.

While grassroots environmental organizing efforts were successful in halting both of these high-profile mining projects, the mining companies did not suffer major defeats. In both cases, the companies were ensured access to other government lands of comparable mineral worth. Nevertheless, the mining industry is not used to the kind of grassroots environmental organizing that stopped the New World and Windy Craggy projects. The permitting of new mines in sensitive areas where local residents place a high value upon a clean environment continues to be a major social problem for the mining industry in most advanced capitalist nations.[2]

The kind of resistance that occurred in Yellowstone and British Columbia is indicative both of the organizing skills of people whose livelihood and culture are threatened and the failures of corporate strategies in seeking to buy off local communities in secret negotiations with elected officials.

While images of devastated landscapes from Appalachian coal strip mining became part of the national environmental consciousness in the 1960s, the far more extensive damage from unregulated hard-rock mining of metals like gold, silver, copper, and uranium has only come to be defined as a major environmental health problem quite recently. In 1989, for example, the U.S. Bureau of Mines reported that such mining has contaminated more than 12,000 miles of rivers and streams and 180,000 acres of lakes and reservoirs in the United States.[3] At least 60 of the 1,381 sites now on the U.S. Superfund hazardous waste cleanup list are former mineral operations.[4] The largest Superfund site is a former copper and silver mining and smelting area where pollutants have migrated 130 miles along Montana's Clark Fork River and contaminated a land area one-fifth the size of Rhode Island.

Even though the hard-rock mining industry generates about the same amount of hazardous waste as all other industries combined, Congress specifically exempted mining wastes from regulation as hazardous waste in the Resource Conservation and Recovery Act (RCRA) of 1976. Moreover, unlike most manufacturing industries, the U.S. mining industry is not required to report its toxic emissions to state and federal regulators.[5]

Special exemptions enjoyed by the hard-rock mining industry have allowed them to avoid public scrutiny and the widespread public opposition that has characterized the siting of hazardous waste facilities in the wake of Love Canal and other toxic contamination disasters.[6] However, as the higher-grade mineral deposits are exhausted and new mining ventures exploit lower-grade ores in more fragile environments, the conflicts between mineral extraction activities and environmental protection become more visible and more likely to generate grassroots opposition movements.[7] This is nowhere more evident than in the attempts to transform large portions of northern Wisconsin into a new mining district.

A New Mining District in Northern Wisconsin?

Beginning with the discovery of the Flambeau copper-gold sulfide deposit in 1968, northern Wisconsin became an attractive target for mineral exploration. The relatively modest 1.9-million-ton Flambeau deposit was soon overshadowed by the 55-million-ton zinc-copper sulfide deposit discovered by Exxon Minerals in 1976. By the early 1980s, as many as fifteen multinational mining corporations were exploring the state for mineral deposits.

Despite several attempts by powerful mining corporations like Kennecott Copper and Exxon Minerals, only one mine has actually been constructed. Grassroots citizen, tribal, environmental, and sport-fishing groups have blocked mining projects in Ladysmith, Crandon, and Lynne. Faced with a series of embarrassing defeats, the mining industry, in cooperation with the state, initiated a variety of strategies to overcome grassroots environmental resistance to new mining projects. These strategies have included the following, among others: (1) legislative initiatives to thwart local democratic control, (2) legal challenges to local zoning authority, (3) mass media campaigns, and (4) attacks on tribal sovereignty.

In his survey of the global antienvironmental movement, Andrew Rowell argues that "the intensity of the corporate counterattack against a burgeoning environmental opposition has been so powerful that in countries like America, it has, at best, derailed, at most, destroyed democracy itself."[8] While this statement may sound like an exaggeration, for many Wisconsin rural communities that have had some degree of success in opposing new mining projects, this is an all too accurate characterization of the erosion of democracy.

Legislative Initiatives to Thwart Local Democratic Control

Before mining companies can receive permits to mine in Wisconsin, they must have the approval of local units of government—a major obstacle for both Kennecott Copper and Exxon Minerals. A decade after withdrawing from the Flambeau project, Kennecott reevaluated the project and discovered that the copper lode was an extraordinarily rich deposit. In 1987 Kennecott reactivated its mine application for a scaled-down version of the defeated project.

However, the company could not meet the tough environmental requirements contained in county zoning ordinances and thus could not get a state permit. In one of Kennecott's "issue papers," the company identified "a small vocal opposition group" whose concerns about mining impacts could be "neutralized" if local leaders and company officials could negotiate a "local agreement" addressing some of these concerns.[9]

To avoid what the company called "onerous local approvals," a Kennecott official drafted the so-called "local agreement" law, which allows mining companies to negotiate a local agreement in lieu of zoning permission. Such negotiations are confined to elected officials. The bill was attached to a budget bill and passed without public hearings or debate in 1988. Shortly thereafter, Rusk County gave in to the mining company's threat to sue for "deprivation of economic use of its property" and signed a local agreement before the Wisconsin Department of Natural Resources (DNR) had even issued an environmental impact statement (EIS).[10]

Nine years after withdrawing from the Crandon project, Exxon and Canada-based Rio Algom formed the Crandon Mining Company (CMC) and resurrected plans to extract 55 million tons of zinc-copper sulfide ore at the headwaters of the Wolf River in northeastern Wisconsin. Shortly thereafter, CMC began closed-door negotiations with local units of government to secure advance permission for the mine through a local agreement. Citizens in the town of Nashville objected to the closed-door negotiations, but their protests were ignored.

Seeking to prevent their town board from giving advance permission for the proposed mine, 230 out of 301 Nashville voters petitioned for a special town meeting. Citizens wanted to vote on whether the town should enter into a local agreement with CMC before all the issues surrounding the mine had been discussed in a master hearing.

Among the major problems the citizens had with the local agreement was the attempt to exempt the mining company from all town zoning ordinances, regulations, and laws and to limit the powers of local government and the courts to prohibit mining.[11] The agreement also gives final township approval for the disposal of all wastes associated with the project. Over its lifetime, the mine would generate an estimated 44 million tons of wastes.

Exxon used the full extent of its financial and political power to get the local agreements approved. They financed full-page ads in local newspapers, bombarded residents with radio ads, and bussed in mine supporters to voice support for the local agreement at the Nashville township hearings. In December 1996, the town board voted to approve the local agreement with Exxon.

Five critics of the local agreement then filed petitions to run for town of Nashville positions in the coming April 1997 election. In February 1997, the Forest County chapter of the Wisconsin Resources Protection Council filed a lawsuit accusing the town board and the Crandon Mining Company of holding more than a dozen illegal closed meetings to develop a local agreement. In announcing the lawsuit, one of the plaintiffs and a candidate for town chairman noted that it was being brought

as a class action on behalf of all citizens whose right to speak out and be heard by their elected officials has been ignored. It is brought on behalf of all residents and tribal members who live and work in the Wolf River and Wisconsin River watersheds in harms way of the potential havoc that this mine may cause and whose rights to clean air and water have been forgotten. These Local Agreements were hammered out in secret, behind closed doors. They are weak and ineffective. They do not protect the citizens of Nashville, Forest County, and the state or protect the rights of tribal members of the Native American nations who live in the two watersheds which will be directly affected by these Agreements. . . . We can't let our communities be sacrificed by corporate greed or let "feel good" television commericals, paid for by Exxon, cause us to forget what is right.[12]

In the April 1997 local election, four out of the five town board members were voted out of office.

Legal Challenges to Local Zoning Authority

The power of large, multinational mining corporations to threaten lawsuits against small rural townships who dare to withhold permission for exploration and mining can be very intimidating. While Exxon–Rio Algom was meeting with the Nashville town board to develop a local agreement for a zinc-copper sulfide mine at the headwaters of the Wolf River, BHP Minerals International, Australia's largest company, was applying for a conditional use permit to conduct mineral exploration and drilling not far from Exxon's proposed Wolf River mine. If both Exxon and BHP were to proceed with their mining plans, the township would be totally surrounded by metallic sulfide mines.

At the public hearing on BHP's application before the Nashville zoning committee, local residents testified that there were no examples of successful sulfide mine reclamation anywhere. They argued that the permit should be denied because the cumulative impacts of exploration and mining had not been identified, the use was not consistent with the development pattern in the town's land use plan, and did not meet the health and welfare concerns of community residents as expressed in a public opinion survey.[13]

Shortly after the zoning committee voted to deny BHP's exploratory drilling request, the company filed a lawsuit against the township and the members of the zoning board, threatening each member of the zoning committee with confiscation of their property and/or wages. The lawsuit accused the zoning board of voting against BHP's application "for reasons that were unrelated to the Ordinance standards." BHP also alleged that the committee's denial was "arbitrary and capricious and takes BHP's property without just compensation in violation of the United State's and Wisconsin Constitutions."[14] The town's board of adjustment then overturned the zoning committee's decision and gave BHP permission to drill. The company dropped the lawsuit.

Mass Media Campaigns

In 1994 Roper Research conducted a survey of how the public perceived various industries. Mining came in last after tobacco.[15] While the mining industry attempts to portray the mine permitting process as a purely technical and scientific process, the industry cannot mine without public approval. A major component of the backlash against environmentalism in the United States and Canada consists of the mass media campaigns conducted with the help of professional public relations firms.[16] In 1990 U.S. businesses spent an estimated $500 million on hiring the services of antienvironmental public relations (PR) professionals and on "greenwashing" their corporate image.[17]

The former chairman and chief executive officer of Freeport-McMoRan Copper and Gold Company, one of the most notorious mine polluters, encouraged his industry colleagues to establish organizations to coordinate and implement "image-enhancement programs for mining."[18]

The immediate impetus to Exxon's media campaign in Wisconsin was the "Save Our Clean Waters Speaking Tour," organized by the Wolf Watershed Educational Project and the Midwest Treaty Network, which built upon previous efforts of grassroots environmental groups, sport-fishing groups and Native American nations. When Exxon announced its plan to divert mine wastewater into the Wisconsin River, the speaking tour expanded to include cities and towns along that river. This opened up a whole new constituency that had not previously been concerned with the project. After each community event, organizers left behind a core of grassroots supporters who carried on the work of coalition building and community action.

Exxon accused mine opponents of spreading misinformation about the project without specifically identifying a single example. Exxon's full-page ads emphasized that "The Wisconsin Department of Natural Resources [DNR] cannot approve a mine that will threaten public safety, harm the environment, or be bad for the local economy."[19] Exxon's tactic of using the Wisconsin DNR to reassure the public was problematic because the pro-mining Governor had just eliminated the public intervenor's office, the state's environmental watchdog, and transformed the DNR into a political patronage agency. All of the state's environmental, conservation, and sport-fishing organizations opposed these moves. Both actions severely undermined public confidence in the state's ability and commitment to protect clean water and public health and safety from the risks of mine pollution. As in the case of hazardous waste facilities, public distrust of government regulatory agencies fueled the local opposition to the siting of these facilities.[20]

The speaking tour was also designed to build public support for legislative passage of a sulfide mining moratorium bill that would prohibit the opening of a new mine in a sulfide ore body until a similar mine had been operated for 10 years elsewhere and closed for 10 years without pollution from acid mine drainage. By focusing public discussion and debate on the problem of acid mine drainage, mine opponents were able to shift the discussion from the issue of *mine production,* which leaves the state, to the issue of *mine waste,* which remains in the local community and may have long-term and serious effects on both the environment and the health of local populations.

A study by the Institute for Environmental Studies at the University of Wisconsin warned that "the potential for damage may be so severe as to require perpetual monitoring and maintenance similar to that done by federal authorities with radioactive waste material."[21] The identification of acid mine drainage as the most serious mining pollution problem became a "political icon" for the mining opposition in the same sense that the 55-gallon drum became a political icon for the toxic waste protest movement of the 1980s.[22]

Prior to the Wisconsin senate's vote on the mining moratorium bill, slick television commercials promoted the wonders of modern mining technology and associated it with the warm, fuzzy images of the idealized version of life in the small, northern Wisconsin town of Crandon in Forest County, where Exxon and Rio Algom want to construct a large underground zinc and copper mine.

The ads showed geese flying over a lake, a sparkling stream, and school children. However, the ads did not discuss controversial issues about groundwater and surface water contamination from acid mine drainage and heavy metals, the drawdown of local water supplies in the vicinity of the mine, and the effects of discharging upward of one million gallons of treated mine wastewater into the Wisconsin River every day for 30 years.

Exxon's second television ad featured United Steelworkers of America (USWA) union president Dennis Bosanac of Local 1114 in Milwaukee with the union seal in the background. He says:

Some legislators in Madison want to stop mining. That's like asking over 10,000 working people to stop breathing. Don't they know that thousands and thousands of us work in jobs that depend on mining? Don't they know how important mining has been and will be to Wisconsin? We want to be part of it. Those high paying jobs belong in Wisconsin, not someplace else. . . . Working together for the future. (Crandon Mining Company)

The Wolf Watershed Educational Project issued a press release explaining why the ad was misleading. First, the mining moratorium bill would not ban mining in Wisconsin. Instead, the bill requires that prior to obtaining a permit, the applicant must demonstrate that a similar mine has been operated for 10 years and closed for at least 10 years without pollution from acid mine drainage or heavy metal contamination. Second, if the Crandon mine is not opened, potential jobs will not travel out of state, because the ore cannot be moved. Mining equipment companies will still receive contracts from outside Wisconsin. While Exxon promises 400 permanent high-paying jobs in the Crandon area, there is no assurance that these jobs will not go to already skilled miners who, after six months in Wisconsin, will become "local" residents. The May 1997 decision of the Copper Range Company to withdraw its permit for acid solution mining at the White Pine copper mine in nearby Michigan will provide further incentive for the remaining miners to seek jobs elsewhere.

Third, the Crandon Mining Company (CMC) ad never mentioned that Exxon and Rio Algom, a Canadian mining company, are the co-owners of CMC. This is a significant omission because the ad leads the viewer to believe that the United Steelworkers support the companies behind CMC. Nothing could be further from the truth. The USWA has been in the forefront against Rio Algom on the issue of worker health and safety at the Elliot Lake uranium mines in Canada. As the main union involved in uranium mining at Elliot Lake, the USWA expressed deep concern over the health effects of radiation from the early days of mining. The Ontario Workman's Compensation Board reported in 1969 that sixteen out of twenty deaths of Elliot Lake miners were the result of lung cancer. A USWA survey showed that "Rio Algom had consistently underestimated hazards in virtually every part of the mining complex and mills, by deliberately under-reading radiation levels."[23]

Seeking to prevent passage of the moratorium bill in the Wisconsin assembly, Exxon set up and funded the Coalition for Fair Regulation

(CFR) in the hope of mobilizing other industries, such as the paper mills, which have not yet been part of the mining moratorium battle. The mining industry has been advocating these kinds of broad-based coalitions with timbering, land development, and paper production as a way to increase its political clout on key legislative battles. The CFR steering committee includes the largest mining and mining equipment manufacturers in the world. The attempt to turn workers against environmentalists failed. Several unions, including the steelworkers and the construction workers, passed resolutions in favor of the mining moratorium bill.

Despite an unprecedented media, lobbying, and mass mailing campaign by Exxon, RTZ-Kennecott, and the Wisconsin Manufacturers and Commerce Association, the unprecedented public support for the moratorium bill resulted in legislative approval by an overwhelming margin (29–3). While the legislation does not stop the mine permitting process, it creates environmental standards that the industry will be hard pressed to meet.

Attacks on Tribal Sovereignty

Indian control over air and water quality on reservations is long overdue. Tribal lands were ignored in the original versions of many federal environmental laws of the 1960s and 1970s, including the Clean Air Act and the Clean Water Act. A direct result of this oversight has been the accumulation of mercury and other toxic chemicals in fish and game to levels posing "substantial health, ecological, and cultural risks to a Native American population that relies heavily on local fish and game for subsistence."[24] As a remedy, in the past decade amendments to these laws have been enacted to give tribes the same standing to enforce environmental standards as states. The position of the U.S. Environmental Protection Agency (EPA) is that the amendments to the Clean Water and Clean Air acts are based upon the inherent authority of the tribes to protect the health and welfare of their members. In 1994–95, four Wisconsin Indian tribes asked the EPA for greater regulatory authority over reservation air and water quality.

In 1995–96, the Sokaogon Chippewa, Oneida, and Lac du Flambeau tribes were granted independent authority by the EPA to regulate water quality on their reservations. The Sokaogon Chippewa's reservation, famous for its wild rice, is just a mile downstream from Exxon's proposed Wolf River mine site. Tribal regulatory authority would affect all upstream industrial and municipal facilities, including Exxon's proposed mine. Because Swamp Creek flows into the Sokaogon Chippewa's Rice Lake, the tribe has to give approval for any physical, chemical, or biological activity upstream that might degrade their wild rice beds.[25] The Wisconsin DNR recognizes that the Chippewa's wild rice is "at the center of their identity as a people."[26]

At public hearings on the Sokaogon Chippewa's application for water regulatory authority, local citizens, lake associations, and the Wolf River Watershed Alliance testified in support of the tribe's application. However, the Wisconsin Mining Association warned that tribal water quality authority "could be the most controversial and contentious environmental development affecting the state in decades."[27]

The Sokaogon Chippewa's water regulatory authority is not the only concern of powerful corporate interests. The nearby Forest County Potawatomi Tribe is seeking federal approval of clean air standards that would affect the ability of large industry to pollute the region's air. Exxon estimates that if the mine is built, it will emit about 247 tons of particulates into the air a year. The Potawatomi air quality regulations would only affect facilities that release at least 250 tons of pollutants per year. If Exxon falls under the regulatory cutoff point, the tribe fears that the airborne mining dust, which contains heavy metals like lead, arsenic, and cadmium, will enter the food chain and accumulate in tribal members who consume fish and wildlife from the area.

Two pro-mining northern Wisconsin Republican legislators commented: "As legislators and concerned citizens we stand united in opposing the imposition of obscure provisions in the federal Clean Air Act which deny the citizens of the state due process, violate state sovereignty and threaten the economic stability of many northern Wisconsin counties and communities in northern Michigan."[28] Both the governor and the secretary of the DNR urged the EPA to deny tribal regulatory authority over air and water quality standards.[29]

Meanwhile, a coalition of legislators and business leaders called upon Congress to change the Clean Air Act to disallow tribal authority over clean air standards.[30] Powerful corporate interests are using scare tactics to suggest that Indian sovereignty over reservation resources is an economic threat to small business owners and ordinary citizens, while they ignore the serious potential for long-term damage to the resource and economic base of northern Wisconsin from large-scale mining and waste disposal.[31]

Within a week of EPA approval of Sokaogon Chippewa water quality authority, the Wisconsin attorney general sued the EPA in federal court, demanding that the federal government reverse its decision to let Indian tribes make their own water pollution laws. "All bodies of water in Wisconsin are public and belong to no one, not even an Indian tribe," said James Haney, a spokesman for the Wisconsin justice department. Therefore, according to Haney, it is the state's responsibility, not individuals or tribes, to set and enforce water pollution standards in Wisconsin.[32] In April 1999, the U.S. District Court in Milwaukee dismissed the Wisconsin lawsuit and upheld the tribe's right to establish water-quality standards for its reservation near the proposed Crandon mine.[33] The state is appealing this decision.

Conclusion

As grassroots resistance to environmentally destructive mining activities succeeds in delaying, modifying, or stopping new mining projects, multinational mining corporations and state agencies have defined this resistance as a social problem in the same way that local opposition to the siting of hazardous waste facilities was defined as a problem in the 1980s.[34] Some of the same strategies that were used to overcome local opposition to the siting of hazardous waste facilities have been used to overcome local opposition to new mining projects.

Among the most important of these strategies have been state preemption of local siting authority and the use of financial compensation to offset community costs with more benefits.[35] While preemption removes control over land use from the hands of local opponents, it does not preempt all forms of local opposition. If opponents of unwanted facilities cannot exercise their right to withhold needed zoning or other permit approvals, they will simply use more creative measures. Preemption laws may also encourage opponents of a facility to challenge the legality of these laws. In 1981 the state of Wisconsin adopted a negotiated siting process for hazardous waste facilities after the courts rejected the state's preemption effort.[36] The addition of a negotiation process still did not result in the siting of a single new hazardous waste facility.[37] The attempts of other states to obtain approval for such sites were no more successful.

The strategy of offsetting community costs with benefits, such as additional public services or tax revenues, distracts attention from the potential environmental harm of a facility. The assumption that local opposition could be bought off has been offensive to many involved in the toxic waste protest movement.

Even if communities are willing to give up their health and safety concerns for financial compensation, there is still a fundamental issue of social injustice. If hazardous waste facilities were allocated to those communities most in need of any additional source of income, these facilities would end up in the poorest and most oppressed communities, exacerbating the already serious problem of "environmental racism."[38] (See the chapter on brownfields in this volume.)

In the Wisconsin case of the local agreement negotiated between the town of Nashville and Exxon–Rio Algom, the Mole Lake Sokaogon Chippewa Tribe, the most directly affected, was not even consulted during the negotiations. Since the Mole Lake Tribe stands to be the community most adversely affected by the proposed mine and its toxic waste dump, and since reservation land is held in trust by the federal government, this could be one of the most serious obstacles to federal approval of the mine project. The courts have ruled that federal agencies cannot subordinate Indian interests to other public purposes except when specifically authorized by Congress to do so.[39]

The question remains: Have the corporate and state strategies for overcoming local resistance to new mining projects been any more effective than similar strategies that were developed to neutralize the toxic waste movement of the 1980s? In the case of local opposition to the Kennecott-RTZ copper mine in Ladysmith, the strategy met with some degree of success. The mine was constructed after numerous delays, court challenges, and civil disobedience actions at the mine site. However, this was a relatively small mine by industry standards and the grade of ore was rich enough to allow the company to ship the unprocessed ore to Canada and thereby avoid the construction of a permanent waste disposal site in Ladysmith. While local opposition was highly organized and motivated, statewide opposition to the mine was limited.

The local resistance to Exxon–Rio Algom's Wolf River mine is a different story. While the company has a local agreement for their proposed mine, they also have a full-blown citizen-tribal insurgency that has thrown out a pro-mining town board, replaced it with an anti-mining town board, and filed legal challenges to the local agreement that could effectively halt the project. In contrast to the Ladysmith experience, Exxon–Rio Algom have not been able to restrict the resistance movement to the local area; it is statewide.

This was nowhere more evident than in the extensive grassroots lobbying campaign organized by supporters of the mining moratorium bill. It was hardly coincidental that after assembly approval of the bill, Exxon announced that they were selling out their interest in the Wolf River mine to Rio Algom. Although the company emphasized that the decision was based on general business needs, polls indicated that the majority of Wisconsin citizens were opposed to the project.

While mine opponents cheered Exxon's withdrawal, it is not the end of the battle. However, the willingness of one of the world's largest corporations to walk away from a large mineral project potentially worth more than $4 billion tends to reinforce the conclusion that the environmental, sport-fishing, and native nations coalition that came together in embryonic form during the Ladysmith mine battle has developed into an effective, mature, and broad-based statewide movement.

The movement's strength is due in no small part to the delegitimation of state and industry authority as they try to force their mining agenda upon communities that are increasingly aware of the health risks of metallic sulfide mining. As indicated earlier, this increased awareness was largely a result of the strategic decision of mine opponents to focus public discussion and debate on the *mine waste issue,* and especially the industry's unsolved problem of *acid mine drainage,* which would affect the state's pristine rivers.

Unlike the hazardous waste movement of the 1980s, however, the resistance to mining is not a typical environmental movement. It is a rural-based, multiracial, grassroots rebellion that has forged significant links with an urban, labor, and student constituency. The diversity of this coalition has continually confounded the mining industry and thwarted attempts to isolate the mining opposition from the political mainstream.

Does's Exxon's defeat mean that the mining industry has exhausted its strategies for overcoming local resistance to new mining projects? Not quite; it simply means that we can expect the corporations to shift the conflict from the local and state level to the national and international level. Under provisions of the proposed Multilateral Agreement on Investment (MAI) for example, Wisconsin's mining moratorium law could be considered a form of expropriation, and a foreign mining corporation like Rio Algom could sue for damages.

Regardless of the MAI, mining industry leaders are worried about their ability to contain a growing multiracial environmental resistance movement in Wisconsin. A recent editorial in *North American Mining* warns industry leaders that if they continue to dictate to communities, they will face a "time bomb of socio-economic concerns which demand just as much attention, patience, cost and effort from operators as environmental protection does today." As examples of where the industry has ignored this reality, the editorial points to conflicts "still brewing between mining companies and local peoples in the state of Wisconsin in the United States, Irian Jaya in Indonesia, the provinces and territories of Canada and states and territories in Australia."[40] In all of these places the industry faces similar political coalitions between environmentalists and native peoples.

Notes

1. David Darlington, "Copper Versus Grandeur," *Audubon* 94:4:90 (July–August) 1992.

2. Sharon Prager, "Changing North America's Mind-Set about Mining," *Engineering and Mining Journal* 198(2) (February 1997):36–44.

3. Robert L. P. Kleinman, "Acid Mine Drainage," *Engineering and Mining Journal* 190(7) (July 1989):16i–16n.

4. U.S. Environmental Protection Agency, "Risks Posed by Bevill Wastes" (Washington, D.C.: EPA, F-98-2P4F-S0032 1997 p. 4).

5. John E. Young, "Mining the Earth," Worldwatch Paper No. 109. (Washington, D.C.: Worldwatch Institute, 1992, p. 35).

6. Andrew Szasz, *Ecopopulism: Toxic Waste and the Movement for Environmental Justice* (Minneapolis, Minn.: Univ. of Minnesota Press, 1994); Michael B. Gerrard, *Whose Backyard, Whose Risk: Fear and Fairness in Toxic and Nuclear Waste Siting* (Cambridge, Mass.: MIT Press, 1994).

7. Al Gedicks, *The New Resource Wars: Native and Environmental Struggles Against Multinational Corporations* (Boston: South End Press, 1993, pp. 46–47).

8. Andrew Rowell, *Green Backlash: Global Subversion of the Environmental Movement* (London: Routledge, 1996, p. 69).

9. Kennecott, Issue Paper No. 1: Local Agreement/Local Approvals (April 25, 1988).

10. Ron Seely, "Northern Officials Give State Poor Grades," *Wisconsin State Journal* (March 24, 1991): p. 13A.

11. Although the Wisconsin law has not yet been judged unconstitutional, common law rules generally prohibit municipalities from bargaining away their policy-making powers. See Lawrence S. Bacow and James R. Milkey, "Overcoming Local Opposition to Hazardous Waste Facilities," in *Resolving Locational Conflict,* Robert W. Lake, ed. (New Brunswick, N.J.: Center for Urban Policy Research, 1987, p. 166).

12. Charles Sleeter, Press release on the filing of a lawsuit against the town of Nashville and Crandon Mining Company (February 10, 1997).

13. Mike Monte, "BHP Minerals International Meets with Nashville Zoning," *Pioneer Express* (Crandon, Wis.), April 22, 1996, p. 1.

14. BHP v. Town of Nashville, Complaint filed in State of Wisconsin Circuit Court, Forest County (June 12, 1996, p. 7).

15. Praeger, "Changing North America," p. 37.

16. Panos Institute, "Green or Mean?: Environment and Industry Five Years on from the Earth Summit," Panos Media Briefing No. 24 (London: Panos, June, 1997, p. 13).

17. Joel Bleifuss, "Covering the Earth with Green PR," *PR Watch* 2(1) (1995):1–7.

18. Milton Ward, "Mining and the Environment: In the Long Run We Are All . . . Survivors." Minerals Industry International. *Bulletin of the Institution of Mining and Metallurgy,* No. 1006 (May 1992):33,35.

19. CMC (Crandon Mining Company), Full-page advertisement. *Daily News* (Rhinelander, WI) May 5, 1996, p. 8.

20. Bacow and Milkey, "Overcoming Local Opposition," p. 160; Szasz, *Ecopopulism,* p. 104.

21. Michael D. McNamara, "Metallic Mining in the Lake Superior Region: Perspectives and Projections," Institute for Environmental Studies Report No. 64. (Madison, Wis: Univ. of Wisconsin, 1976, p. 51). The majority of Americans view the mining industry primarily as a source of valuable raw materials rather than as a major contributor to pollution. In a recent poll, 57 percent of the 1,000

people surveyed said that they view mining companies "primarily as producers of valuable resources." Only 21.9 percent said that they viewed the industry primarily as a polluter. See *Engineering and Mining Journal* (1997)198:11:16EEE (November).

22. Szasz, *Ecopopulism,* pp. 63–64.

23. Roger Moody, *Plunder!* (London: Partizans/CAFCA, 1991, p. 127).

24. U.S. Environmental Protection Agency, "Tribes at Risk: The Wisconsin Tribes Comparative Risk Project" (Washington, D.C.: EPA, 1992, p. ix).

25. Don Behm, "2 Tribes Hope to Control Reservations' Water Quality," *Milwaukee Journal,* February 5, 1995, p. 1B.

26. Wisconsin Department of Natural Resources, Final Environmental Impact Statement, Exxon Coal and Minerals Co., Zinc-Copper Mine, Crandon, Wisconsin, Madison, Wisconsin (November 1986).

27. James Buchen, "Delegation of Federal Clean Water Act," *Badger State Miner* (October/November 1995):1.

28. Lorraine Seratti and Tom Ourada, "Air Redesignation Concerns Legislators," *Forest Republican* (May 4, 1995), p. 1.

29. Jeff Mayers and Nathan Seppa, "Thompson Vexed by Tribe's Move." *Wisconsin State Journal* (December 13, 1994), p. 1B.

30. Steven Walters, "Tribes Request Could Jeopardize Current, Future Jobs, Groups Say," *Milwaukee Journal Sentinel,* April 27, 1995, p. 3B.

31. Gedicks, *New Resource Wars,* p. 167.

32. Mike Flaherty, "State Sues Over Indian Water Law," *Wisconsin State Journal* (January 30, 1996), p. 1B.

33. State of Wisconsin v. U.S. EPA and Sokaogon Chippewa Community; U.S. District Court for the Eastern District of Wisconsin; Case No. 96-C-90. April 30, 1999.

34. Szasz, *Ecopopulism,* p. 105.

35. Daniel Mazmanian and David Morell, *Beyond Superfailure: America's Toxics Policy for the 1990's* (Boulder, Colo.: Westview, 1992, p. 109).

36. Ibid., p. 189.

37. Ibid.

38. Szasz, *Ecopopulism,* p. 110. See also Robert Bullard, "Anatomy of Environmental Racism and the Environmental Justice Movement," in *Confronting Environmental Racism: Voices from the Grassroots,* Robert D. Bullard, ed. (Boston: South End Press, 1993).

39. Janet Smith, Comments to the U.S. Army Corps of Engineers from the U.S. Department of the Interior, Fish and Wildlife Service, Green Bay Field Office (November 1994, p. 2).

40. *North American Mining Magazine,* "Smart Mining Companies Emphasize Local Partnerships," 1(9) (November 1997):3.

15

Bearing Witness or Taking Action?: Toxic Tourism and Environmental Justice

Giovanna Di Chiro

Leading a "toxic tour" of Newtown, Georgia, an African American community located 55 miles north of Atlanta, Rose Johnson speaks gravely about the "high rates of throat and mouth cancers, excessive cases of the immune system disease, lupus, and a variety of respiratory ailments" that have afflicted the residents of this small southern town. "Too many people have died" is the message that comes through her portable microphone as she directs the visitors along a somber route. Moving door to door, neighbor to neighbor, Johnson visually maps the terrible trail of pain by respectfully placing black ribbons on the homes of residents who were ill or who had died of cancer, lupus, or from chemically induced heart disease. The several dozen guests participating in this toxic tour, first organized in 1993 and repeated many times since, consisted of city and state officials and members of the National Council of Churches who wanted to experience "firsthand the sharp contrast between the Newtown environment—where the acrid odor of toxic industry and scrap yards presses in on the little park and well-tended homes—and the flourishing green lawns and flowering trees of Longwood Park and the generous houses on the north side of town where most whites live."[1]

Traveling in an old yellow school bus and on foot, visitors to Newtown follow an itinerary charted on a community-generated "toxic" map that traces the location of thirteen industrial sites and numerous hazardous waste generators surrounding the residential area of the town. These maps, which profile the toxic emissions from nearby industrial facilities, are developed through a collaboration between Newtown residents and concerned environmental scientists, epidemiologists, geologists, and legal experts using the federally mandated toxic release inventory (TRI) database. Johnson's toxic profile map helps her navigate this guided tour, which demonstrates the spatial relationships between industrial facilities and serious health problems. Visitors from other cities and towns begin to envision a scene of environmental catastrophe; this is not the usual fare offered up on an ordinary tourist excursion.

This is not an ordinary sightseeing tour, and Rose Johnson is not a typical tour guide. She is a member of the Newtown Florist Club, a community organization founded in the 1950s by local women to generate the funds to provide flowers for funerals and "to care for the sick and comfort the families as they buried their dead."[2] More recently, the club has turned its attention to the health of the entire community and organizes against the environmental injustices perpetrated by numerous toxic polluters that have long imperiled the community. Activists such as Johnson, Faye Bush, and Mozetta Whelchel organize the toxic tours to provide unassailable physical evidence that Newtown residents are suffering from a disproportionate impact of hazardous pollutants. As they lead their tour groups from one site to the next, they enable their visitors to see for themselves what is graphically documented on the toxic profile map—the high density of industrial facilities located in and around exclusively black communities. Guests to Newtown can "sightsee" the close proximity of schools, playgrounds, and houses to toxics-emitting facilities that include the Leece-Neville machine plant, which discharges thousands of pounds of xylene and 1,1,1-trichloroethane every year; the Ralston-Purina and Cargill feed mills, which release heavy metals, pesticides, and sulfuric acid into the local atmosphere; and the ConAgra chicken processing plant, whose emissions include ammonia, arsenic, and zinc.

The final stop on Rose Johnson's tour is the Bethel African Methodist Episcopal church. At the church, tour participants disembark, "feeling the weight of evidence of life in a toxic zone."[3] This is the environmental "reality" for the predominantly African American residents of Newtown—a reality that has been briefly and poignantly experienced on this day by a group of outsiders, most of whom are white. Many visitors attending these uncommon tours avow it is the closest they have ever come to witnessing firsthand the consequences of what activists have long contended is overt environmental racism. The toxic tour has succeeded in illuminating the historically invisible components of the Newtown landscape—the devastating effects of the externalities of industrial development on the land and the people who live on it. Providing the opportunity for this firsthand, or "authentic," experience of environmental injustice partly underlies Johnson's and the club's use of toxic tourism in their strategy for social and environmental change. The act of seeing with one's own eyes is not intended to be the only experience, however. Their aim is to take action to change what the eyes witness.

Ecotourism with a Twist

Tourism, that "essentially modern" practice of constructing self-consciousness by locating oneself at a distance and differentiated from the "other," which some argue is a distinguishing feature of western cultures,[4] currently enjoys a lucrative proliferation in an environmentally aware niche market popularly known as "alternative tourism." Like all tourism, alternative tourism aims to provide direct access into another's world or worldview, to gain a deeper involvement with other peoples and other places. Tourists set out on journeys to faraway places to escape the anxieties and alienation that have accompanied rapid industrialization and urbanization in modern, western societies. Consequently, as theorists of tourism argue, the tourist hopes to respond to this alienation spawned by modernity by reconnecting with nature, particularly "edenic" nature, as a source of inspiration and belonging— to engage in authentic communion with the natural world and with those earth-based peoples who are perceived to have never lost the connection.[5]

Within the alternative tourism industry, like its more traditional predecessor, "authentic" experience of the "other" (in the form of pristine nature as well as native culture) is still the saleable commodity. However, these primarily western, middle- to upper-middle class tourists are seeking a real-world travel experience not simply to gawk, but to ask questions and to find answers: How did we get ourselves into this ecological conundrum, and how can we learn from the environmental knowledge of the indigenous cultures of the world in order to create solutions? The alternative tourism industry exploits these sensibilities, clear signifiers of the modern "environmental imagination,"[6] and presents tourism, not simply as a consumer product, but rather as a cultural practice that can be transformed into an activity serving progressive ends.

Critics of this new trend suggest that an endlessly renewable resource, liberal guilt, fuels the success of alternative tourism, primarily owing to the perception that "people occasionally need to take vacations even while the rainforest is dying."[7] Alternative tour packages promote the idea that you can do something to protect the environment while on vacation.

Of the many forms of alternative tourism, which include culture tourism, science tourism, and archaeotourism, by far the most popular is ecotourism. As defined by the Ecotourism Society (established in 1991), ecotourism is "responsible travel that conserves natural environments and sustains the well-being of local people."[8] The environment is the fastest growing draw in the alternative tourism industry worldwide. Ecotourism, which centers on nature, environmental awareness, and native cultures as objects of desire, promises to be a multi-billion-dollar industry by the year 2000.[9] Ecotourism as "tourism with a message," aims to promote a sense of urgency about the global threat to spectacular nature and indigenous cultures while simultaneously offering up a great vacation. It strives to promote sustainable development in ecologically sensitive and threatened regions of the world by taking advantage of the revolution in the international tourism industry. Compared with other kinds of industrial development, the environmental impacts of tourism development often seem less damaging; supporters of ecotourism tout it as helping to "build healthy relationships" between endangered ecosystems, foreign exchange–wielding tourists, and the economic welfare of impoverished, native peoples.[10]

Clearly, the current trends in alternative tourism identify both the environment and the condition of local people who make their living in particular environments as potentially important tourist draws that not only satisfy industry desires to turn a profit, or tourist desires to partake in an authentic nature experience, but also environmentalist desires to improve the deteriorating condition of the global environment.[11] The ecotour constructs itself as representing the seamless convergence of capitalist political economies with environmental well-being. Ecotourism companies endeavor to offer the consumer a risk-free, enjoyable holiday that satisfies multiple aesthetic, economic, and political needs.

Another alternative form of tourism, also emphasizing the interrelationships between the environment and local cultures, is *toxic* tourism. Toxic tourism signifies ecotourism with extra value added. The toxic tours led by the women activists of the Newtown Florist Club are certainly environmentally based tours, or *ecotours,* but in this case they highlight the not-so-scenic sites scattered on the landscape and cast a very different light on the assertion that a compatible relationship can exist between capitalism as we know it and the health of humans and the environment. On the ecotour, a.k.a. *toxic tour,* of Newtown, Georgia, the environment and the community are intertwined, and the dangers of separating the two are exposed. The "reality" that is uncovered in the toxic tourism experience, as indicated earlier, is often disheartening. Toxic tours are not "feel good" vacations, but does this mean that tour-goers will not go back home feeling committed and inspired to act? As in all forms of ecotourism, toxic tours enjoin people to learn something along the way. What kinds of knowledge and understanding occur when participants return from a toxic tour? What are some of the benefits and problems of toxic tourism as an environmental justice strategy?

In this chapter I argue that the growing popularity of the ecotourism industry has opened up potentially interesting and innovative political spaces for environmental justice organizing. This most recent version of modern tourism culture, one that is laced with a healthy dose of eco-liberalism (in other words, the belief that we can protect the quality of the environment by commodifying and consuming its exotic features), is the historical context in which toxic tourism has emerged. Toxic tourism exploits this historical convergence of the tourism industry with the discourses of sustainable development and environmental consciousness. The proliferation of ecotourism has provided grassroots environmental groups with new openings for resistance and new spaces for imaginative forms of environmental justice activism.

This chapter addresses three versions of toxic tourism as a political strategy used by a variety of social and environmental justice organizations. First, I discuss the toxic tour as a political demonstration. This type of tour directs participants on a predesignated protest march with the intention of engaging them in political opposition against the particular toxic offenders who have business offices or corporate buildings located at each tour stop along the way.

Second is the toxic tour as a "reality tour" offering an alternative vacation. In this version, the tourist pays for an organized tour package that includes the costs of transportation, meals, and lodging and most often includes traveling to distant locales. Tour-goers experience a theme-driven, guided tour of critical sites (polluting facilities, chemical spills, toxics-damaged neighborhoods, working conditions at factories, sludge-filled rivers, etc.) that are indicators of a specific socio-environmental problem emphasized on the tour. The reality tour also encourages tourists to talk with local people who are battling these injustices.

The third kind of toxic tour is the model used by the women of the Newtown Florist Club. These tours, undertaken by communities who are themselves suffering from the effects of environmental injustice, seek to build supportive connections with members of other communities or with sympathetic allies in government and the private sector. These tours lead visitors through the toxic sites in the community, but also invite participants to "sightsee" some of the more positive features of the community environs.

Toxic Tours as Political Protest: The Toxic Links Coalition

Harking back to the freedom marches that inspirited the civil rights movement of the 1960s, environmental justice organizations have used toxic marches as political action for many years. The series of Great Louisiana Toxic Marches, starting in 1988, were massive demonstrations that sent protesters along Louisiana's "Cancer Alley" between New Orleans and Baton Rouge along the Mississippi River. Moving systematically from one oil refinery, chemical manufacturer, or incinerator to the next, brandishing posters, signs, and resonant voices, marchers demanded an end to what they considered corporate America's toxic assault on people of color and low-income communities.[12]

More recently, a coalition of over thirty environmental, health, and human rights organizations in the San Francisco Bay area adopted this "political protest" form of toxic tourism. The Toxic Links Coalition

(TLC), founded in 1994, is an alliance of a wide range of community organizations dedicated to demonstrating the links between the accelerating decline in public health and communities' exposure to the increasing presence of environmental toxins. A brochure published by the coalition states that the TLC "views cancer and other environmentally linked diseases and disorders as human rights abuses, not individual medical problems; targets companies with irresponsible practices for handling carcinogenic waste and products; demands accountability from corporate and agricultural polluters; and works against environmental racism." To inaugurate the establishment of the coalition, the TLC organized its first annual toxic tour, otherwise known as the Cancer Industry Awareness Tour, of the San Francisco financial district.

The TLC's Cancer Industry Awareness toxic tours are conducted during October, the month chosen years earlier in a large public relations campaign underwritten by the pharmaceutical and chemical titan, Zeneca, as Breast Cancer Awareness Month. As a counter-PR action, the TLC renamed the campaign, Cancer Industry Awareness Month, to expose what they argue is a "public relations scam to promote mammography and to obscure the real causes of cancer."[13] With participants chanting the organization's overarching slogan, "stop cancer where it starts," and accompanied by a police escort, the tour begins in front of the corporate headquarters of Chevron, identified as "part of the cancer industry" and whose East Bay oil refinery and pesticide manufacturing plant in North Richmond receive regular citations from the U.S. Environmental Protection Agency (EPA) for hazardous emissions violations.

The tour is always held on a workday during lunch time for maximum visibility and to accommodate working people willing to relinquish their lunch hour to march. Participants receive information at its start by activists from a variety of community, health, and environmental justice organizations, that corporations like Zeneca do not have cancer prevention at heart. Zeneca manufactures known carcinogenic chemicals such as the organochlorines used in pesticides and herbicides, and a variety of high-priced drugs used to treat cancer. Anger at this obvious conflict of interest spurs the more than 250 demonstrators on to the next tour stop—the federal building and the local offices of the EPA and U.S. Senator Diane Feinstein.

At every stop, tour-goers hear speeches given by local activists who describe continuing struggles with the target corporation, agency, or official. At the federal building, for example, activists from Clean Water Action and the Political Ecology Group attest to the EPA's misguided plans to commission an unlined, low-level nuclear waste dump a few feet from a major aquifer in the Ward Valley, and Senator Feinstein's support of immigrant and welfare reform bills, which activists argue scapegoat poor people as the cause of environmental problems, rather than focusing on the corporations responsible for pollution and environmental decay.

The tour moves briskly along the course, stopping at the headquarters of a major public relations firm, Burson-Marsteller, which represents large chemical and pesticide companies that produce well-documented carcinogens. On the October 1996 tour, Dr. Jan Kirsch, an oncologist and member of Physicians for Social Responsibility confronted an imagined, unnamed executive of Burson-Marsteller, inviting him to "come make my rounds with me in the hospital" to see for himself the effects of his company's business activities.[14] Activists sometimes appeal to the target corporations to change their ways and join the struggle for environmental justice. Such target groups include Bechtel, the nuclear power plant builder; or the local utilities provider, Pacific Gas and Electric; or the American Cancer Society, which has never taken a stand on the environment and instead promotes "early detection," drug treatment, and lifestyle changes rather than identifying environmental and industrial changes to prevent cancer.

Along the route, participants are also given some positive news—that one company, San Francisco Energy, has been dropped from the tour because, in response to the protests on the previous year's (1995) toxic tour, the company decided to halt production of a new plant in the primarily African American community of Bayview-Hunter's Point, a community already heavily affected by toxics that has unusually high levels of breast cancer in women under the age of fifty. The unity of many diverse environmental and health organizations in the area has helped to stop the new plant and has succeeded in eliminating one of the tour's target sites. At the Pacific Stock Exchange, the last tour stop, a tour organizer urges passers-by to "come out with us next year, even if it means giving up your lunch break—the life you save may be your own."[15]

The TLC's annual toxic tours are well organized political demonstrations that are not about "touring the actual toxic sites, but rather the places where, symbolically, the decisions are made" (personal communication from Jan Arnold, member of the TLC, June 13, 1998). As a highly visible and noisy political protest staged in the generally staid, gray-suited quarters of San Francisco's financial district, the toxic tours hope to attract "someone who just happens to be in the area and who is grabbed by our message."[16] The TLC's toxic tours are not technically a form of ecotourism because the participants are already converts, quite well informed, and strongly persuaded about the seriousness of the issues. They attend the tour, not to have a discovery experience, but to make a strong, visible political statement opposing certain corporate and government actions while assembling at the doorstep of the corporate headquarters of American business.

Vacations with a Message for a Changing World: Global Exchange's Reality Tours

After attending one of the twenty possible "reality tour" excursions sponsored by the San Francisco–based nonprofit organization, Global Exchange, tourists comment on their experiences:

This was a real fact-finding tour. I could watch, touch, listen and converse using my eyes, ears, hands and skin.[17]

I've been all over the world, yet this trip affected me more than any other. It was hands-on, not theory or ideology . . . but talking with and looking into the faces of real people.[18]

This has been a real antidote to the hopelessness I sometimes feel.[19]

Hearing the responses of participants who had traveled on one of a number of reality tours dealing with "provocative themes such as the environment, art, culture, revolution, and peace" in the United States and abroad, one gets the sense that "reality" or authenticity was truly the tour's object of desire. Tour-goers experienced, not the customary scenes of spectacular nature that draw millions of people into the ecotourism industry each year, but rather the locales not usually listed on the chamber of commerce's "must see" registry. The itinerary of a California-based reality tour might include a stop at an East Bay oil refinery to see its impact on neighboring communities; or a strawberry farm in central California to learn about the impact of the soil fumigant methyl bromide on the farm workers, the consumers, and the environment; or one of 700 foreign-owned maquiladoras located on the California-Mexico border to see the miserable and poisonous working and living conditions of thousands of Mexican workers.

These tour sites unveil the hidden and unsightly parts of the landscape, transforming them into real social and environmental effects, not side effects, of global capitalism. However, after a tourist encounter, the "guest" goes back home (typically to one of the affluent urban centers of California), and the question arises: What exactly is the reality or authentic experience that is being traded on here? Or, as an organizer of the border environmental justice campaign from the San Diego–based Environmental Health Coalition, César Luna, asks, "What do you plan to do with this information?"[20]

Reality tours are the brainchild of Medea Benjamin, the co-director of Global Exchange, a "human rights, education, and action center" founded in 1988 "dedicated to promoting people-to-people ties between the North and the South" and to "encouraging the U.S. government and international institutions to move toward policies of democratic and sustainable development."[21] The organization achieves its objectives through numerous program areas, including public education efforts and the fair trade program, which helps to generate income for artisans in over thirty-seven countries. Yet the reality tours are unique. The common theme woven through all the organization's projects, and especially meaningful in the tours, is building personal contact through productive partnerships at local, national, and international scales. North American travelers are given the opportunity to encounter and learn from "people you'd never get to meet on your own."[22] An urban professional from Los Angeles might get the chance to meet with government officials in Haiti, indigenous women community organizers in Chiapas, Mexico, or factory workers in a free-trade zone at the U.S.-Mexico border. Flexibility, sensitivity, and an open mind are prerequisites for attending a tour (in addition to a price tag ranging from $30 to a few thousand dollars). According to tour leaders, using a socially responsible form of ecotourism to build link-ages based on understanding between people from different cultures, classes, and geographical locations is a first step toward constructing a strong grassroots base to demand more socially just and sustainable mod-els of development.

Originally the reality tours aimed to promote socially responsible travel and "true understanding" by North American tourists of the environmen-tal predicaments of peoples in the poor countries of the South. Global Exchange's reality tours were a direct challenge to the burgeoning market for "offbeat, educational vacations" reflected in the emergence of eco-tourism in the late 1980s. The reality tours, organized around a specific theme, seek to raise important social, economic, and environmental issues faced by communities in different countries. For example, some of the theme-based tours include Women in Chiapas: Indigenous Women and Self-determination (Mexico); Land Struggles, Indigenous Rights, and Culture in Brazil; and Grassroots Development from a Gandhian Perspec-tive (India). In contrast to the "feel-good" orientation of most ecotours, reality tours do not necessarily paint a pretty picture of the exotic lands and the daily realities of the people who live in the vicinity of the tour stops. Reality tour-goers learn about the residual miseries of the global-ization of markets and international trade by talking directly to farm la-borers, indigenous agriculturalists, or women union organizers in a foreign-owned factory. As one commentator remarked, these tourists "paid for a vacation they knew would depress them, in the hopes that it would also inspire them."[23]

By 1997, organizers of the reality tours determined that North Ameri-cans, especially those of the middle and upper classes, "could use a reality check in their own backyard." Instead of heading off to distant lands to seek out bona fide socioenvironmental problems, Global Exchange orga-nizers crafted the new tour program, "Exploring California." The critical issue, according to Benjamin, was whether these locally based toxic tour packages would attract the alternative tourism market. Surprisingly, she remarks, people are willing to "pay money to see toxic sludge."[24] In Au-gust 1997, on a week-long excursion to the northern part of the state, Global Exchange launched a domestic initiative with the Redwoods Tour: Sustainable Forests and Communities in Humboldt County. Offering hiking, swimming, and camping in the Humboldt Redwoods State Park *and* informational meetings with local residents, native communities, government officials, forestry scientists, loggers, and mill owners, participants are given an intensive immersion experience into the continuing struggle to save both the 60,000-acre privately owned Headwaters forest and the livelihoods of local people.

Later in the fall of 1997, Global Exchange conducted the first Ecological Faultlines Tour. This is a one-day toxic tour with a sliding-scale fee ranging from $50 to $100, and focusing on a range of environmental justice struggles taking place in the San Francisco Bay area.[25] Cosponsored by the Urban Habitat Program, this toxic tour introduces participants to grassroots activists from San Francisco, Oakland, Richmond, and points in between, who are on the front line working for environmental justice.[26] It aims to build partnerships and understanding between environmentalists concerned with more traditional issues, such as deforestation and ecological preservation, and those who focus on issues associated with environmental justice, such as battling against the siting of hazardous facilities in communities of color and working to eliminate the production of and dependence on the toxic substances that contribute to the acceleration of environmental health problems worldwide.

The Ecological Faultlines tour attempted to traverse the gaps, or faultlines, that have historically separated well-meaning, sympathetic, mainstream environmentalists from environmental justice activists and began to forge a mutual understanding that may lead to collaborative action. Lisa Russ, one of the organizers of the Faultlines tour, explains that people were drawn to the tour because they are curious about what's going on since they live in the area and want to be informed. Many wanted to know what the environmental justice community groups were doing and saw the tour as an efficient way of "surveying" the local scene.[27] However, curiosity can sometimes mutate into prurient interest. Russ elaborates:

It's our responsibility as tour organizers not just to feed the interest in "what's the ugly side of a corporation" but to show the solution side. The "toxic tour" label is a problem because of the risk of sensationalism. One of the problems and risks of doing a political tour like this is that you can sell difference and you can sell inequity and that's gross, you know, "come see the poor people, come see the toxic sludge." But that just reinforces the idea that "I am privileged and I don't have to live like that"; it doesn't have a useful political message. So, we're trying to do things in a different way. We have to resist sensationalism and insensitivity at every level—like bringing outsiders to a community to talk with people who are in a hard situation.[28]

Russ argues that the toxics-damaged communities who agree to participate as one of the tour stops are not interested in the "white liberal guilt" response of pity. Organizers of the Faultlines toxic tour initially canvassed numerous community organizations in the area to determine their interest in being involved and to identify their views on the critical issues, major goals, and primary target audiences for the tour. Working closely with the activists from the communities who are suffering environmental injustices before they may become further exploited as stops on a sightseeing adventure is an effort "to challenge top-down tourism."[29]

How did the community organizations benefit from their involvement in the Faultlines toxic tour? Russ says they wanted to reach funders and to connect with technical experts who could be of assistance to them; some of these goals were met. Six people from the EPA's National Environmental Justice Advisory Council (NEJAC) attended the tour; some groups were promised funding; some groups received news coverage of their recent campaigns. Russ does not unequivocally support this kind of tour, nor does she necessarily advocate the continued involvement of Global Exchange in this specific project; the experience was mixed. The grassroots groups, she explains, "are thinking they want to organize these tours on their own."[30]

The Redwoods tour was more successful than the Faultlines tour, Russ argues, because structurally it is a week long and involves a small group that camps together every night and is "immersed in the experience rather than taken on a survey experience." The environmental justice issues are handled on a deeper level by studying a small community. And perhaps most important:

The person who leads the trip is a local organizer in Humboldt County. He feels that educating people about the issues is important. Community groups there feel that we've got to get people up here, to show them what this is all about. You've got to tell a more detailed story than you get in the media. The Redwoods tour is a more locally generated initiative to get the information out about these issues. Last year the participants made direct changes in their own lives and became activists in many ways—by attending rallies, by educating their own friends, by changing their world view. That's the kind of thing we'd like to see happen.[31]

When organized under the rubric of ecotourism or even socially responsible tourism by an organization or company that is outside of the impacted communities themselves, toxic tours run the risk of appropriating the control over self-representation and self-determination that are central components of a marginalized community's political empowerment. "Top-down tourism" reproduces the legacy of the imperialist history of tourism and the colonizing tourist-host relationship. Russ's reflections on some of the thorny issues that are inherently raised in an organizing effort like the Ecological Faultlines tour, even with the best intentions of creating partnerships and political allies, suggests a need for careful planning in the early stages of organizing. Another toxic tour organized by Global Exchange's Exploring California program, this time examining the theme of trade, the environment, and social justice at the Mexican border, raises other critical issues that emerge when organizations choose tourism as a political vehicle.

In April 1998, Global Exchange hosted a customized reality tour called "What do you mean, Free Trade?" for a high school student group from the Sacred Heart Preparatory School in Atherton, an elite community bordering Silicon Valley. On this toxic tour (which cost around $450), students are taken on a six-day spring break expedition that investigates the environmental and human rights effects of international trade agreements like the North American Free Trade Agreement (NAFTA) on local communities close to home and those living in border towns in Mexico. Many of these students come from families working in the high-tech computer industry who can afford to pay the $13,000 per year tuition of the preparatory school. The tour boasts a full schedule that introduces them to a host of community organizations like the Silicon Valley Toxics Coalition, a nearby group that educates them about the dismal labor practices and lenient environmental standards pervasive in the computer industry.

Loaded into tour vans, the students travel to Los Angeles, where they meet with labor and immigrant rights groups who highlight the connections between unemployment, poverty, air pollution in East L.A., and international economic developments like free trade agreements. The tour introduces the northern California students to young people in the southern part of the state and in Tijuana and provides historical and political economic explanations for their different economic futures. In Tijuana, the students meet with maquiladora workers, some of whom are teenagers like themselves, and see the dilapidated shacks built from wooden shipping pallets and discarded plastic sheets housing several families, and smell the acrid odors of sulfuric acid fumes or sewage runoff from one or another industrial plant. This is a vacation? Other Sacred Heart Prep students, who in all likelihood went skiing at Lake Tahoe for spring break, may wonder what their classmates were thinking. Some of the tour guides wonder the same thing.

What exactly are these young people expected to do with the information and with the experiences that they get on this toxic tour? César Luna, who meets with students on one of the stops at the Environmental Health Coalition in San Diego, wonders about what kinds of messages register with young people who attend these "free trade" tours. Overall, he states, students return very positive responses; "they get to see some 'reality' with their own eyes and it is very challenging for them."[32] Other students, however, become overwhelmed with emotions like guilt and respond by expressing gratitude for their good fortune to live in America—as one Sacred Heart student replied, "It's not [my] fault [I] live in an upper-middle-class family in a first-world country."[33]

Luna worries that the students who pass through his stop on the tour "sometimes do feel impotent and a little guilty about coming from a better place, and my response is first that they shouldn't feel guilty because it's *not* their fault," but by becoming proactive they have a lot of opportunity to create change in the face of these injustices. The question for Luna is, do the toxic tours provide such opportunities for action? The response is one of ambivalence. On the one hand

you bring general awareness—the general awareness that no, NAFTA is not as beneficial as people think it is, there are people suffering, there are people with problems relating to the "free trade" that NAFTA has engendered. The tour changes their visions, or at least the stereotypes with respect to how things are on the border—so that in itself is education and it's positive.[34]

On the other hand, Luna continues, in a practical sense, considering the time and resources these "general public" toxic tours consume for grassroots organizations, the payback is not great.

If you had a tour of scientists or a tour of media people or a tour of elected officials, chances are that you may be able to get some results. What is the best way that we can benefit this community by bringing people here—so that they not only get more familiar with the issues but so that they get more proactive?[35]

In line with the general principle that ecotours (according to the Ecotourism Society) and toxic tours (according to tour organizers like Global Exchange) must benefit both tourists and "host" communities, both Russ and Luna pose the question: Who benefits from these tours? The consciousness raising and educational gains are clear to most tour leaders (although some have worried that the tour format itself unavoidably creates the conditions that promote voyeurism and objectification of local peoples like a "trip to the zoo or Disneyland"[36]). What are the effects on the host communities? Sometimes just being there and offering a sympathetic ear helps. One young man from Tijuana reflected on his discussions with the high school students on the free trade tour: "More than anything, the moral support you give us is important. . . . that means a lot, because no one else listens."[37] Some tour leaders and community activists wonder if, in fact, that is enough.

As Russ noted above, upon reflection, she would have changed the organizing process of the toxic tours from the outset. Such tours, she argues, must be initiated by the communities themselves. Luna agrees. The terms of these regularly organized toxic tours "have to be much more in the interests of the local communities." To these ends, Luna argues, all participating organizations and the people they will be visiting must be involved at the planning stage deciding what is best for them and how bringing outsiders in would expedite change. Referring to the toxic tour as a political "tool" for change, he also warns that it must be used cautiously, with sensitivity, and strategically. It should not "turn the communities into parks" that tour groups are perennially passing through. According to Luna, the issues of who controls the agenda and whose interests are being served are paramount. He argues that visitors have a certain responsibility to give something back to the community. Political involvement is not a risk-free enterprise. However, ecotours (and some toxic ones) are intended to be.

What is the best way to assess the effectiveness of these formalized toxic tours directed by organizations that are not aligned with any particular community? Earlier, Russ, troubled about the impact of the Ecological Faultlines tour, wondered whether her organization should be involved at all. Luna, in response to the young high school students' anxieties about whether they would need to live in a third world country to help change the world, explained that staying put has its advantages:

You may be more valuable in the U.S. And you may be able to make a bigger change if you stay home than if you come here. Your skills and know-how may be more limited here than in your own environment. I think they got the picture.[38]

Toxic Tours and Community Action: Practicing Environmental Justice

Environmental justice organizations want to build political constituencies that will advance their cause. Toxic tours are one educational and political tool that many community organizations use to tell their stories and promote face-to-face, personal connections. Activists hope that these brief snapshots of the daily realities of people struggling with the devastation of toxic pollution will move beyond "opening one's eyes" and raising general awareness, and toward positive action to help change the unjust social and environmental conditions they have witnessed.

Communities of color and low-income white communities in the United States who are organizing against environmental racism and the disproportionate impact of hazardous pollution have long resisted being (mis)represented by others and have insisted on "speaking for ourselves." Visitors on the Newtown toxic tour walked side by side with local residents and activists and listened to and saw, if only momentarily, what it's like to live in an "industrial fallout zone."[39]

Environmental justice organizations speak for themselves on the many community-led toxic tours offered around the country. These locally initiated tours are designed to show outside groups, whether schoolchildren, official EPA delegations, or skeptical environmental scientists, the complex realities of the community's circumstances. This may include scenes of a deteriorating housing stock with unsafe lead levels or a Texaco oil refinery abutting an elementary school attended by low-income African American and Latino students. However, guests would also be conveyed to local sites that make the community distinctive, such as an architectural landmark or a thriving community garden, in addition to meeting with community groups who are actively working to improve their environments.

Two prominent environmental justice organizations in the San Francisco Bay Area, the West County Toxics Coalition (WCTC) in Richmond and People United for a Better Oakland (PUEBLO) in Oakland, participated as site visits on Global Exchange's October 30, 1997 Ecological Faultlines tour. Henry Clark from WCTC discusses his organization's interest in leading toxic tours:

We've been doing the tour for many, many years. We've taken classes from UC Berkeley, from Stanford, Berkeley High, Richmond High, community groups that request the tour, so we've done the tour for many different organizations over the years. We show them the area, and the work that we're involved in so that they get some first hand sense of the issues and the territory that the WCTC organizes in. The tours certainly are helpful in terms of environmental education and whatever contacts we might make that may be useful in our work.[40]

Some tour-goers have returned as interns, volunteers, or doctoral researchers, all of whom help in the day-to-day activities of the organization. For example, student researchers use their skills to help document the evidence of noncompliance with environmental emissions regulations that are consistently charged against the Chevron oil refinery and the Chevron Partho-Chemical Company, the two major toxic facilities located in this primarily African American community. Although Clark applauds the benefits of the WCTC bringing outsiders into the community to see what's happening with their own eyes or listening to residents with their own ears, he argues that just witnessing the outcomes of environmental injustice in communities of color is not always productive in bringing about change. Rather, he advocates good, old-fashioned organizing to generate grassroots pressure on a particular agency, company, or decision-maker. It is community organizing, Clark argues, that compels a local company "to sign a good neighbor agreement, or to reduce pollution risks by installing pollution prevention equipment, or to close down older parts of the refinery. It's usually not a result of any recognition that there is a problem or that the company itself expresses willingness to invest resources in addressing it."[41]

Karleen Lloyd from PUEBLO explains that her organization does not have time to conduct tours, but is "available when other groups put on a tour." PUEBLO's interest in participating in the Ecological Faultlines tour was "to make our campaign more visible, to educate people about what's going on, how different communities have different struggles, and in some cases to build allies with people."[42] Also a grassroots community organization, PUEBLO emphasizes person-to-person organizing styles, but deploys a variety of innovative technologies to aid in publicizing their environmental justice projects.

Lloyd speaks enthusiastically about the organization's new geographical information system (GIS) program, which will enhance their ability to demonstrate visually on multilayered maps the relationship between the spatial distribution of hazardous industries, such as the Integrated Environmental Systems (IES) medical waste incinerator and Owens-Brockway glass manufacturer, and a variety of social and economic factors. These socioeconomic factors might include the industries' proximity to residential areas, the incidences and clustering of health disorders caused by exposure to industrial toxins, and racial and ethnic demographic profiles of neighborhoods bordering toxic facilities. Lloyd argues that people can be persuaded to act on the strength of seeing the facts in front of them, whether they are depicted on a multilayered GIS map or in a site visit during a community-led toxic tour:

While maps like anything else, can be disputed, we are a very visual society. Showing maps to members in the community is more persuasive than just telling them what's here or what's there. The map tells a story with a picture and I think that is more persuasive.[43]

In New York City, West Harlem Environmental Action (WHEA) organizes toxic tours for school groups, community residents, and other environmental organizations interested in knowing more about environmental justice activities. However, since these tours take time and resources away from the already overburdened staff, WHEA is selective about who they will take on a tour. Strategically, according to organizer Cecil Corbin-Marks, the most effective target audiences to lead on a toxic tour are funders, reporters, and community members.[44] These groups can potentially give back to the organization.

The WHEA tour moves through northern Manhattan, including Harlem, central Harlem, west Harlem, and Washington Heights, a region with a population of more than half a million people. A diverse community of

African Americans, Latinos, and Asians inhabits a part of the city that is also home to a variety of toxic sites, including the North River sewage treatment plant, marine transfer stations handling tons of garbage, bus depots leaking diesel fuel, and numerous brownfields. Corbin-Marks, who leads the two-hour tour, explains that WHEA is not interested in presenting an image of the upper Manhattan communities as victims who should be pitied, and so the tour also stops at a variety of cultural landmarks such as the Apollo Theater and the Hope Community Garden.

One of the central uses of the toxic tour by environmental justice groups across the country is to educate and lobby government officials, journalists, and decision-makers from the public and private sectors. PUEBLO and the WCTC were two of the community organizations that participated in the EPA's NEJAC regional meeting in the Bay Area in June 1998. Part of the meeting was a day-long site tour in which the NEJAC members traveled by bus to different organizations to see and hear about the environmental issues of concern to their communities. In July 1998, the NEJAC delegation went to the Philadelphia area for a toxic ecotour of Chester, a city of primarily African American residents who have recently sued the Pennsylvania Department of Environmental Protection, charging "environmental racism" for granting five low and high-level waste site permits in this community of 40,000 poor and low-income people of color. One piece of evidence supporting the accusation is that "the Chester sites have a total permitted capacity of 2 million tons of waste per year, compared with 1,400 tons per year for sites in the rest of Delaware County."[45]

Local organizer Zulene Mayfield, from the environmental justice group, Chester Residents Concerned for Quality Living (CRCQL), led the delegates along a charted route that followed the "smells and noise" from multiple waste incinerators dispersed throughout the neighborhoods and a steady stream of tractor-trailers with full cargoes of medical waste, garbage, and toxic materials from all over the state heading toward the various dump sites. For a stunning visual contrast, Mayfield took the group of official visitors to Swarthmore, a neighboring affluent college community overflowing with elegant homes and well-manicured, verdant landscapes. This sensory comparison enhances the visitors' opportunity to experience life in an industrial sacrifice zone.

Clark's, Lloyd's, and Mayfield's interest in hosting these toxic tours is to provide indisputable physical and visual evidence to convince the federal government to act quickly to eliminate environmental injustice in its environmental siting and regulatory policies. Their modified "ecotours" just may contribute to shaping a new set of environmental practices that are founded on principles of social and economic justice.

Conclusion: Ecotourism Gets a Reality Check

The explosion of general interest in preserving the environment and its human inhabitants, as evidenced in the popularity of ecotourism worldwide, presents activists with an opportune climate to develop creative environmental justice strategies. Exploiting modern, western societies' appetite for touring to gain direct, unmediated experience of the world, environmental justice organizations make use of the tour as one of the many potentially effective social change strategies in their organizing toolkits.

Modern tourism brings together visitors and natives in an encounter of authenticity. Alternative tourism, under the rubric of ecotourism, also sells an experience of authentic contact, but offers something a little different—an encounter with the genuine threat of environmental problems.[46] Ecotourism strives to provide that true-to-life experience for tourists so they will appreciate the purity and value of nature and native cultures before modernity sets in and destroys them. Ecotourism urges the sightseer to "see things as they are" and to protect them so they stay that way. It is a practice of promoting the preservation of the naturalness of a prelapsarian world.

Toxic tourism, on the other hand, although not in the business of profiteering, still trades on the hallmarks of modern tourism—the power of the visual, the physicality of experience. Toxic tours challenge conventional tourism practices but also appropriate some of its methods. The toxic tour, however, impels the tour-goer not to preserve the reality they are confronted with, but to transform it. The toxic tour brings the visitor face to face with the hidden externalities of industrial society, rather than with the imagined purity and innocence of the world before modernity. Toxic tourism can be understood as a species of ecotourism, what I have called alternative ecotourism, even though it is not a money-making venture because of its focus on the relationship between environmental degradation and social problems and the belief that firsthand experience may result in environmental action. Both toxic tourism and ecotourism challenge the remoteness of the "touristic gaze," a kind of museumlike looking from afar, and instead seek to create the conditions for an interactive form of sightseeing. Both genres of environmental tourism aim to present an experience of natural and social environments "as they really are."

Playing on ecotourism's claims to market an authentic nature/culture experience, which often is replete with fun-filled frolics in tropical waterfalls and strenuous treks along mountain slopes accompanied by native guides, toxic tourism suggests that there is another, less attractive, but compelling, side to that ecoreality. For the leaders of the many toxic tours available across the country, it is precisely that other view that must be critically illuminated and transformed if in the future those sites of spectacular nature and traditional cultures so desired by ecotourists will continue to exist.

Different environmental justice organizations employ toxic tourism in different ways and for varied ends. The toxic tours led by the TLC, which take the form of political demonstrations, target specific corporations and officials who, they argue, are directly responsible for toxic pollution or for covering it up. These highly visible political protests bring to public awareness the hypocrisies in, for example, the Chevron corporation's "People Do" PR campaign about the environment. On a toxic tour that targets the corporate office of Chevron in San Francisco, marchers and bystanders learn much more about the oil giant's hidden messages and activities that directly affect the lives of real people and real environments.

As one of the hallmarks of the social change strategy of the organization, Global Exchange, reality tours directly challenge the sugar coating that whitewashes some ecotours. Reality tours seek to provide paying tourists with an alternative travel experience; one that exposes them to the destructive effects of global political economic and environmental policies and practices, as well as to the diverse groups of people on the front lines struggling to bring about change. As a strategy that adopts the format of tourism, and since the organization is not aligned with any specific community, the reality tours can run the risk of reproducing the voyeurism (albeit unintended) and "othering" that characterizes the un-self-reflective versions of ecotourism.

Some tour organizers are rethinking the role of reality tours to reduce the threat of turning the communities on the tour stops into toxic "theme parks." Community activists and reality tour leaders alike argue that the interests of the people suffering environmental injustice must be at the center of organizing efforts in the initial planning stages of a toxic tour.

Although they consume time and resources, for most community groups toxic tours are important political tools that help peel off the blinders that limit fuller understanding of the causes of environmental problems and their potential solutions. Moreover, they present to the visitors on the tour a sense of the anger and spirit that galvanize the people of the community and that push them to continue the fight for environmental justice. It is hoped that greater understanding will lead to informed action. As Faye Bush and Rose Johnson from the Newtown Florist Club urge, "we invite you to join us and share our struggle."[47]

Notes

1. Ellen Griffith Spears, *The Newtown Story: One Community's Fight for Environmental Justice* (Gainesville, Ga.: Center for Democratic Renewal and the Newtown Florist Club, 1998, p. 2).

2. Ibid., p. 4.

3. Ibid., p. 6.

4. Dean MacCannell, *The Tourist: A New Theory of the Leisure Class* (New York: Schocken Books, 1989, p. 15).

5. For analyses of the history of tourism in modern, western cultures see MacCannell, ibid.; Peter Burns and Andrew Holden, *Tourism: A New Perspective* (London: Prentice-Hall, 1995); Simon Abram, Jacqueline Waldren, and Donald Macleod (eds.), *Tourists and Tourism: Identifying with People and Places* (Oxford: Berg Publishers, 1997); John Jakle, *The Tourist: Travel in Twentieth Century North America* (Lincoln: Univ. of Nebraska Press, 1985).

6. See the essay by Darnovsky in this volume for an interesting treatment of the "potentials and pitfalls" of the environmental imagination as it is expressed in middle-class America.

7. Adam J. Freedman, "Ecotopia," *National Review* (December 11, 1995):38.

8. "How Green Can You Get?" *Economist* 346(8050) (January 10, 1998):s16.

9. Sheila Polson, "Touring Light: The Ecotourism Society Sets the Standards," *E Magazine* 9(1) (1998):44; Deborah Mclaren, *Rethinking Tourism and Ecotravel* (West Hartford, Conn.: Kumarian Press, 1998).

10. David King and William Stewart, "Ecotourism and Commodification: Protecting People and Places," *Biodiversity and Conservation* 5(3) (1996):293–305.

11. For critiques of the relationship between the goals and practice of ecotourism, see Joe Bandy, "Managing the Other of Nature: Sustainability, Spectacle, and Global Regimes of Capital in Ecotourism," *Public Culture* 8(3) (1996):539–66; Ian Munt, "Eco-tourism or Ego-tourism?" *Race & Class* 36(1) (1994):49–60; J. Giannecchini, "Ecotourism: New Partners, New Relationships," *Conservation Biology* 7(2) (1993):429–32.

12. Pat Bryant, "Toxics and Racial Justice," *Social Policy* 19 (Summer 1989): 48–52.

13. "Toxic Links Coalition—Stop Cancer Where it Starts," video produced by Bradley Angell, Judy Brady, and Karen Susag, Options 2000 International (San Francisco, 1997).

14. Ibid.

15. Ibid.

16. Ibid.

17. Comments printed on a publicity brochure from participants attending various "reality tours" sponsored by Global Exchange, 2017 Mission St. #303, San Francisco, Calif. 94110.

18. Ibid.

19. Stephanie Simon, "Exploring California: Tourism with a Message," *Los Angeles Times,* February 15, 1998.

20. Denise Dowling, "Let's Go on a Guilt Trip," *New York Times Magazine,* June 7, 1998, p. 84.

21. From Global Exchange's promotional literature.

22. Ibid.

23. Simon, "Exploring California."

24. Ibid.

25. "Ecological faultlines" are defined on the fliers advertising the tour as "1. a relationship between social and biological organisms to their environment and to each other at the bioregional borders. 2. a tour of the relationship of groups of people to their environments and to each other where the bioregion breaks along ecology, race, color, gender, and the disproportionate impact."

26. The environmental justice organizations that participated in the October 1997 Ecological Faultlines tour included The Chinese Progressive Association, People Organizing to Demand Environmental Rights (PODER), San Francisco League of Urban Gardeners, Southeast Alliance for Environmental Justice, West

County Toxics Coalition (WCTC), People United for a Better Oakland (PUEBLO), and Center for Third World Organizing (CTWO).

27. Author's interview with Lisa Russ, Global Exchange, San Francisco, California (April 27, 1998).

28. Russ interview (1998).

29. Ibid.

30. Ibid.

31. Ibid.

32. Author's phone interview with César Luna, Environmental Health Coalition, San Diego, California (June 22, 1998).

33. Dowling, "Let's Go", p. 86.

34. Luna interview (1998).

35. Ibid.

36. Ibid.

37. Simon, "Exploring California."

38. Luna interview (1998).

39. Spears, *Newtown Story,* p. 2.

40. Author's interview with Henry Clark, West County Toxics Coalition, Richmond, California (May 1, 1998).

41. Clark interview (1998).

42. Author's interview with Karleen Lloyd, PUEBLO, Oakland, California (April 30, 1998).

43. Lloyd interview (1998).

44. Author's phone interview with Cecil Corbin-Marks, West Harlem Environmental Action, New York (June 23, 1998).

45. Paul Nussbaum, "On the Bus for Eco-tour of Chester," *Philadelphia Inquirer,* July 28, 1998.

46. Here I am referring to the ecotourism companies that venture to create an environmentally friendly industry that enjoins its customers to save the environment. There are many variants of ecotourism, however. As profit-making enterprises, ecotourism businesses market different experiences. Some ecotourism companies (referred to as "ecotourism lite" by critics), in a ploy to cash in on the worldwide craze for nature adventures, are fairly light on the environmental message component of the trip and heavy on expensive adventures like scuba diving and jungle safaris. Other companies shamelessly advertise themselves as ecofriendly by encouraging customers to "come see the Amazon, the greatest show on earth, before it closes. Who knows how long it can hold out against the ravages of man" (Epirotiki Cruises brochure).

47. Spears, *Newtown Story,* p. 1.

16

For Generations Yet to Come: Junebug Productions' Environmental Justice Project

John O'Neal

By and by, when the morning comes;
When all the saints of God are gathering home.
We will tell the stories of how we've overcome and we'll understand it better by and by!
—Traditional hymn

Junebug Productions is a nonprofit cultural organization based in New Orleans. We produce, present, and help develop storytelling, theater, music, dance, and other artistic work that represents, supports, and encourages working-class African American leadership in rural and urban communities in the Black Belt South. We serve as a resource for the development of cultural programming for leaders in communities that are working to improve the quality and character of life for themselves and others who are similarly oppressed and exploited. Established in 1980, Junebug Productions succeeded the Free Southern Theater, which was founded in 1963 as part of the cultural wing of the civil rights movement.

Our strategy is to create partnerships with people in community organizations, activists, artists, scholars, and others who are working to create healthy, clean communities for all to live, work, and play in. Of all our partners in the effort, two have been especially important: the Gulf Coast Tenants Association (GCTA) and the Peoples' Institute for Survival and Beyond.

GCTA is a coalition of tenants' organizations that has identified environmental racism as a major priority. From Gulf Coast we learned about the fast-growing movement among people of color for environmental justice. Their "Southern Manifesto," developed for the Southern Regional

Summit of People of Color for Environmental Justice in New Orleans in 1992, was a major tool for clarifying our stance on the issues. The Peoples' Institute trains and provides technical assistance to community organizers. Its analysis focuses on the centrality of racism to American culture and the need for an organized effort to undo the damage it has caused as we work to build a more just society. From this dialogue we developed the following working definitions:

• "environment" is every place where people live, work, and play
• "toxic" includes all substances that can injure or kill people by direct and indirect contact
• "racism" refers to the use of historically established racial biases and/ or prejudices to make social, economic, and political policies that penalize one or more racial, ethnic, or cultural entity.

While the problems of environmental toxicity and racism are global, we in southern Louisiana are confronted by particularly virulent forms of them. New Orleans lies near the end of a 150-mile stretch of the Mississippi River dubbed "Cancer Alley." Over 130 major petrochemical plants, grain elevators, medical waste incinerators, solid waste landfills, and other industrial operations poison the water, air, soil, workplaces, neighborhoods, schools, homes, and other elements essential to human existence.

Many of those who defend the environmental status quo[1] argue that the tradeoff is a good one. They say that we must accept a measure of poison to have jobs. The effort of the Shintech Corporation to place a new plastics manufacturing concern alongside the Mississippi River in Convent, Louisiana, offers a classic example of this. Convent, a small town of approximately 4,000 permanent residents perched on the river about 30 miles above New Orleans, is 80 percent black in a 50 percent black parish (county). Advocates for environmental justice, including the Sierra Club and our partners the GCTA, helped residents of the area organize a strong multiracial coalition there in an intense effort to keep the factory out.

Shintech, like other global corporations who own such properties in Louisiana, is a formidable foe. These corporations benefit from huge tax waivers, even as they export the lion's share of the profits from the state (and often from the country). Top management flies in for periodic inspections, while most of the folk who live in the communities where plants are sited are still locked in poverty.

Most of those who suggest that it's expedient to accept a measure of poison for the economic benefits that follow do not live in the toxic environments that they create. At Shintech, the most toxic emissions are usually saved until the weekends and onsite managers and imported professionals lead the rush to get out of town before five o'clock on Friday afternoon so they won't have to see, smell, or suffer the worst of the emissions.

However, African American people, like many others, have finally begun to make connections between the pollution of their environment, injury to their health, and racism. The stakes are high and the issues are complex and will take a long time to resolve, but the work has begun. This situation provides an ideal context for using theater and the arts to achieve change.

Artists as Activists

Our work at Junebug Productions rests on the idea that our duty as citizens takes priority over our license as artists. The artist cannot stand apart from his or her community. Whether the judgments of our work are critical or encouraging, we are obliged to seek those judgments from the people who make up the constituency that we understand ourselves to be a part of. This must be an active process of building mutually beneficial relationships among artists, organizers, activists, and the organizations with whom they work. The goal is for the artists to get access to the people, the ideas, the history, the information, and the stories that will go into their art. By forming direct partnerships, artists can also gain access to the kind of critical judgment that will help to ensure that the ultimate product of their efforts will correspond to the interests and needs of the desired audience. The benefit for the activists, organizers, and their organizations is that they will gain valuable educational resources that they can use to build support for those things that are positive, expose those things that are dangerous, illuminate important principles, and develop critical thinking skills among their constituencies.

Some question our efforts on the grounds that our ideas are more political than artistic. In response to this we argue that politics and art are complementary, not opposing terms. A major part of each generation's work in improving the world is to create a way for those who come after to continue the struggle. Art and culture are essential parts of the process by which people develop a public consciousness of self. Art that helps us come to terms with our deepest and most urgent concerns in a given historical, social, and political context is the art to which we ultimately assign the greatest value. Only such art retains its value over the long term. Whether it is tragic or comic; whether it is seen, heard, touched, smelled, tasted, or perceived in some other way, the primary measure of the value of art is how much it contributes to our capacity to meet the most important moments of our lives. As such, art is one of the principal means by which we make sense of our experiences. The art that we produce is like a letter that we leave for the generations yet to come.

Because Junebug Productions is based in New Orleans, we support and encourage the particular efforts of those in our region who are working to improve the quality of their lives by alleviating the difficulties caused by environmental racism here. Similar problems bedevil every community. The story of the people in New Orleans' Agriculture Street Landfill neighborhood is not unlike the story of people who once lived in Love Canal. Chicano workers obliged to work with toxic chemicals in fruit and vegetable fields in the Southwest, people living in the shadows of atomic energy plants in the Tennessee Valley, all have the same problem. In Institute, West Virginia, there was a chemical plant owned by the same company and *exactly* like the plant that leaked poisonous gas and killed 2100 people in Bhopal, India.[2] People in one place can benefit from what other struggling communities do and the stories they tell. The success of the effort to achieve environmental justice in one place is accelerated by the victories or retarded by the defeats in another place.

The Environmental Justice Project

In 1991 we began work on an environmental justice project that led to the production of a major festival in New Orleans in May 1998. We decided on *environmental racism*[3] as a theme because the most valuable art draws on our most important concerns for its themes.

The major objectives for our environmental justice project were to:

• host a major arts event that would increase public awareness of the problems of environmental toxicity in "Cancer Alley," the exceptional suffering borne by people of color, and the emblematic character of the situation

• strengthen support for community organizations working to deal with the problems of environmental toxicity and racism

• position the arts as useful tools for oppressed and exploited people to use in their efforts to improve the quality and character of their lives

• strengthen the arts organizations, artists, community organizations, and activists who participate

• strengthen the network between the participating persons and organizations

• increase the number of opportunities for effective engagement among people who wish to help with the work of:

—educating ourselves and our communities

—developing viable alternatives for individuals, communities, businesses, and government

—putting economic pressure on businesses and political pressure on government

The main concerns at the beginning were to flesh out the concept of the project, develop the program plan and budget, and identify the community partners, the artists, and the funding partners with whom we would work to create the project. Because of the scope of the problem and the ambitiousness of our vision, preparations for the festival were spread over several years.

Funding is always a problem for the arts, particularly those that are innovative or controversial. The conservative landslide of the early 1990s resulted in cuts in our own funding and the loss of a group of cultural organizers who were to be assigned to each community participating in the project.

Racism intensifies the funding problem for black theater and art. Ultimately, those who stand to benefit from the work that would be done must take the financial responsibility for it. While there is an extensive history of black *popular* entertainment, the tradition of black theater is relatively sparse and highly speculative at best.

Theater depends on small, intimate audiences. It can, therefore, deal with more complex ideas. As a result, the per-unit cost of theater tends to be high and the ideas that it expresses tend to reflect the views and ideas of the people who pay to watch it. Under the best of circumstances it is not likely to make much money. Popular entertainment, such as film and TV, depends on large numbers of relatively inexpensive tickets. They express relatively simple ideas simply. Those who are inclined to resist such strictures are forced to find alternative, innovative ways to survive on the social and economic fringes. Such art tends, therefore, to rely on institutionalized subsidies from the leadership of those who feel they are well served by that art.

Unfortunately that sector of the African American community with the means to provide support, the black middle class, is so insecure in its own standing that it cannot yet offer a reliable base for the support of art aimed at the interests and concerns of the black working class.

The comfortable middle-income sector is where the most likely audience for black theater comes from. However, the black elite are now able to move out of low-income areas to the suburbs with their economic peers, thus depriving most African American communities of important human social and intellectual resources. The irony of this is heightened by the realization that the advances of the present generation of the black middle class are largely the result of the 1960s civil rights movement, which was powered primarily by the passion and struggle of southern, rural, working-class African Americans and later by dispossessed black urban youth. In an action typical of transitional groups, the ebony class of "the newly not poor" has developed a brand of social myopia that leads them to pay more attention to where they'd like to go than where they have been.

The black middle class, therefore, tends to imitate the values and outlook of the white American middle class. Our environmental justice project, the construction of partnerships with community organizations and artists who share our concerns, is one response to this problem.

Still undaunted by a lack of funds, we launched the project with a planning conference in 1993. We convened artists, activists, scholars, and organizations whose work suggested that they shared our concerns.

The festival itself includes formal and informal art of different forms, styles, and genres, practiced by people from different communities, with different skill levels, at different stages of development. We honored the stories of people who have been/are/will be dealing with the problems of environmental racism in a segment of the festival we called "envirotales." We honored the stories of those whose only victory is that they live to fight another day, as well as those who have won significant victories. Not only were peoples' stories used as the basis for original art work that was presented at the festival, we provided opportunities for people to tell their own stories.

One highlight of the festival was the performace of "The Ladies of Treme," a group of six women who live in Treme, the oldest continuously lived-in majority African American community in the United States. The Performing Arts Center, the most sumptuous facility for the performing arts in New Orleans, sits in the center of an unfinished park in Treme near the municipal auditorium and the presumptive home of jazz music, Congo Square. The park itself is a study in racism and irony. It was carved out of Treme in stages over the strenuous objections of the community. In 1939 the New Orleans Municipal Auditorium was built after the city displaced residents of six square blocks of the historic community. Thirty years later, the Performing Arts Center and the park were added to the site, consuming twelve more blocks of Treme. As a concession to the residents of Treme and the African American heritage of the community, the park was named for Louis Armstrong and the Performing Arts Center was named for Mahalia Jackson.

The richness of their history and the resilience of their culture was evident in the stories the Ladies of Treme told, each of which ended with the observation, "but it's not like that anymore." The Ladies of Treme concluded their presentation by lifting traditionally decorated umbrellas above their fancy African frocks; one of them cast aside her walking cane and they did a line procession to the wings of the theater.

The best example of the implementation of the idea of collaboration between artists and community was the relationship between the Carpetbag Theatre of Knoxville, Tennessee, and the New Orleans Americorps

Project. Linda Paris Bailey and others from Carpetbag Theatre made several trips to New Orleans over the term of the festival project for planning conferences, residencies, etc. Between 1991 and 1998, when the festival occurred, Junebug Productions presented each of the artistic partners at least once to afford the artists and the community groups the opportunity to work together. On prior visits to New Orleans[4] the Carpetbag artists had formed good relationships with their community partners, provided workshops for them, and collected a number of stories, many of which became the basis of an extraordinary play called "Nothin' Nice."

The play centers on the coming of age of a young man whose mother is suffering from cancer contracted as the result of environmental racism in a New Orleans community. The young man discovers that he has no alternative but to fight for a clean, healthy place for his family. Rich with music and the detail afforded by the true-to-life stories gleaned from the Americorps youth, "Nothin' Nice" is a powerful work of art that we intend to present again in New Orleans and take with us to Convent next year.

The performance of the play in New Orleans was greatly enhanced by four youths whose stories were reflected in the play and were shared as enviro-tales before the Carpetbag performance. It was a remarkable experience to discover the relationship between the true stories and the fictional play as the performance unfolded.

The biggest "hit" of the festival was an original production of the "Lower 9th Ward Positive Outreach Leaders," a group of African American youths organized and directed by one of the outstanding local actors, Kathy Randels. The performance was presented in a natural amphitheater formed by the junction of the levees restraining the Industrial Canal and the Mississippi River.

The Industrial Canal is at the crux of a long-standing battle over issues fueled by environmental racism. For more than 30 years, the powers that be have sought to expand the canal to facilitate more water traffic through it, despite the clearly stated objections of the lower 9th ward community, which is overwhelmingly black. The Positive Outreach Leaders worked on their project for more than two years. They interviewed elders in the neighborhood and researched the public record to identify important events in the history of the lower 9th ward and used that information as the basis for a very creative presentation that includes music, movement, huge puppets, and some fine acting. Several lower 9 residents and others from the city at large came several times to participate in the performance. The Positive Outreach Leaders were themselves deeply affected by their work and want to keep their group together and continue in community-based arts work. This is another of the groups with which we intend to carry the concept of the festival forward.

An interesting complement to the 9th ward youth group was a joint performance by the Young Guardians of the Flame, a New Orleans Mardi Gras Indian club headed by young chief Brian Nelson, and youth from Idiwan An Chawe of the Zuni Pueblo in New Mexico. Idiwan An Chawe, in another collaborative performance with Roadside Theater called "Corn Mountain/Pine Mountain," opened the festival. "Corn Mountain/ Pine Mountain" compares and contrasts traditional methods of story telling and land tenure from within the Appalachian and the Zuni cultures.

Junebug Productions' contribution to the festival was a reading of a new play called "Just Like Poison Ivy." Like the Carpetbag Theatre piece mentioned earlier, "Poison Ivy" is inspired by the situation of people from the Agriculture Street Landfill Association (a part of the Gulf Coast Tenants Association) and their stories. We also drew on the extraordinary history of the Treme community and stories from a story circle based at the St. Augustine Church. "Poison Ivy" is a love story about a family rich in the culture and history of the Treme community that is almost completely ruined by the toxicity of a place like the Agriculture Street Landfill before they learn that they have to fight for environmental justice to defend their dreams, their lives, and their right to hold property.

With help from the New Orleans Video Access Center and the New Orleans Film Society, we showed videos and films every afternoon and evening for the ten days of the festival. The most outstanding experience was the video documentation of environmental racism made by media artist Branda Miller. Miller has produced a moving documentary on environmental racism in three communities, including "Cancer Alley." She provides all of her unedited data on a CD-ROM and a Web site so that others may use her primary source data to construct their own documentaries. I believe that she has opened a door to the future in her work. Several other film and video presentations were available in a video room throughout the festival.

Each of the artists and some of the scholars and activists who participated in the festival were encouraged to carry out educational activities throughout the community while they were there. One of the most exciting of these was a presentation at Hope House, a facility that provides services to residents of the St. Thomas/Irish Channel neighborhood who have been displaced by community development activities. Singer-storyteller Paula Larke, who participated in the Carpetbag Theatre production entranced the neighborhood audience for over an hour. They were so engaged that they took her to lunch at a neighborhood senior center and invited her back for a six-week residency the following summer. We also mounted an exhibit that introduced several young artists who had work related to our theme hung in the main gallery of the Contemporary Arts Center—the central site of the festival. The atrium of the center hosted a "green expo" during the festival also. The festival closed with a moving celebratory ritual open to all who had participated in it.

To be sure, we have not solved the problems of environmental racism in our community. We did, however, make significant strides forward: We built and expanded on meaningful networks of artists and community activists. We heightened awareness of the problem of environmental racism, especially in our geographic area. Several of our community organization partners increased their use of the arts in their own work. Several new works were created, some of which are available for touring, together and separately. Finally, Junebug Productions has grown clearer and stronger in its own work as a result of the exercise.

The greatest lesson that we learned about doing art work that supports the interests of oppressed and exploited people was the importance of a community organizer. From now on at least one full-time community organizer will be a high priority in our work. We also learned that documentation is important. There is no point in doing good work in the world if no one knows about it.

Conclusion

Regardless of the current boom in certain sectors of the economy, comparative measures show that the living conditions of the great majority of African Americans have grown worse over the last half of the twentieth century. Now and for considerable time to come some of the most important moments of our lives will fall along the axis created by the age-old struggle for racial justice and the relatively new problem of an environment made toxic by the shortsightedness of powerful corporations. Because of institutionalized racist behaviors, the situation is especially threatening to blacks and other people of color. If we fail to act effectively to solve these problems now, we will burden coming generations with even greater, perhaps even fatal, difficulties.

Notes

1. This includes the head of the state chapter of the NAACP, Ernest Johnson, and several of his business associates who have made common cause with Louisiana's Governor M.J. Foster Jr. Around the same time that Mr. Johnson announced his support of the Shintech plant plan in Convent, Governor Foster announced a plan to support one of Johnson's business ventures. The national office of the NAACP has challenged Johnson's actions on the basis of environmental racism.

2. A videotape documentary about conditions in the plant in Institute, West Virginia, has been made by Ann Jackson and is distributed by Appalshop film and video of Whitesburg, Kentucky. It was run as part of the film and video portion of the Echo Fest.

3. The term "environmental racism" became current in 1987 with the publication of a study by the United Church of Christ's Commission for Racial Justice Now, which established definitively that the correlation between toxicity and race is greater even than the correlation between toxicity and poverty. The findings of that historic study have been independently verified several times since.

4. Over the five years leading up to the festival, Junebug Productions presented Carpetbag Theatre, Roadside Theater, and the Urban Bush Women two to four times in performance and/or residencies, which gave them the opportunity to research the issues in our area and to meet with local groups to build partnerships on which new productions could be based.

17

Media Art and Activism: A Model for Collective Action

Branda Miller

This chapter shares the goals, struggles, and processes that created *Witness to the Future,* a model for media art and activism that includes an experimental 50-minute documentary video and a CD-ROM with updatable World Wide Web links. Produced in collaboration with community residents who were transformed into environmental activists, the project focused on three sites in the United States where environmental toxins have affected the health of hundreds of thousands of people: Hanford Nuclear Reservation in Washington State, the San Joaquin Valley in California, and "Cancer Alley" in Louisiana. *Witness to the Future* portrays the struggles of individuals who discover the power of collective action; it contains diverse underrepresented voices, including workers, mothers, whistleblowers, and farmworkers; and African, Native, and Mexican Americans.

A fundamental feature of effectively integrating media art and activism is the participation of the subject in the production process. This requires transferring the knowledge of media production and distribution directly to the subjects of the video. Through collaboration in writing and directing, in determining content and structure, and in evaluating and shaping the way the final product is presented as a tool for organizing collective action, the subjects can control their own representation. This method provides opportunities for reappropriating the framework of debate about the most basic social and political issues.

A central premise of this method reflects the theory of participatory action-research or PAR, a technique practiced in many parts of the world.[1] Incorporating research, education, and sociopolitical action, PAR views the development of knowledge as the integration of experience and commitment. In contrast to the subject-object relationship in traditional academic research, a transforming subject-subject relationship between researcher or communicator and people occurs as they become animators of their own knowledge through authentic participation. For those who are victimized by structures in which they feel powerless, as is so often the case for communities battling large industries and government institutions over environmental and health concerns, the self-awareness and self-confidence resulting from this process develops self-knowledge, action, and power. Community residents, after all, are the experts and spokespeople for their communities.

In an essay on research and education, Paulo Freire wrote: "If I perceive the reality as the dialectical relationship between subject and object, then I have to use methods for investigation which involve the people of the area being studied as researchers: they should take part in the investigation themselves and not serve as the passive objects of the study."[2] With Freire's observation in mind, participatory research seeks to break down the distinction between the researchers and the researched and the subjects and objects of that search by having the people themselves participate in the attainment and creation of knowledge. In the process, research is viewed not only as a means of creating knowledge but also as a tool for the education and development of consciousness and for mobilization for action.[3]

Grassroots organizations' use of media and new information technologies for participatory research and organizing action occurs at a moment when many communication technologies are becoming increasingly monopolized by corporate interests. Seemingly new opportunities and freedoms for direct reporting occur simultaneously with invasions of privacy and the potential for controlling access to information. Until the mass media began to dominate as the main sources of knowledge about the world, many peoples created their identities and culture by telling their stories to themselves, without intermediaries to interpret them. Today, many grassroots organizations, fighting for the health of their communities, increasingly seek ways to reappropriate their history and communicate their experience to wider publics without a filter.

Power, Politics, and Objectivity

As they seek the truth and speak out concerning environmental threats surrounding them, many people's positions are often discredited because the very depth of their personal experience implies that their views cannot not stand up to standards of "objectivity" and scientific analysis, which actually reinforce the institutional power they are attempting to criticize. These notions of objectivity deserve further analysis.

Each night on TV, and in newspapers and magazines, "experts" who seem to represent a broad spectrum of points of view debate together. Yet these experts are almost always affiliated with dominant institutions of power; they are government representatives, corporate executives, professors, and scientists. Rarely does a person who is not in a position of power have a platform in the media, unless they are cast in the role of "victim," where their sharing of vivid tales of murder, disaster, or sex scandal reinforces the entertainment value and therefore the commercial value of broadcast media.

Credentials of expertise, which are typically organized for commerce, are assumed to ensure the "authenticity" of coverage of an issue. Most experts that we hear from are white men, usually financially well-off. Powerful corporations such as GE, Westinghouse, and Disney own the major networks who present the experts. These realities are not readily apparent to the viewer. With practically a monopoly in communications through multi-billion dollar corporate mergers, information delivered for profit, no matter how many hundreds of channels become available, provides little space for other voices and independent visions, especially those who speak out against corporate plundering of the environment. The corporations that hold major interests in industries responsible for polluting the earth themselves have interests in and exert great control over the mass communications industry as well.

The mass media have extraordinary power to depict a subject, designing the frame in which viewers approach issues and ideas. Selecting who will be heard, how they will be represented, and what messages will be delivered, these representations shape our identity, our perception, and even our memory.

Challenging these false constructs of "objectivity" and "authenticity," *Witness to the Future* shifts the dialogue to the real experts, who defy media stereotypes: self-educated citizens fighting for the safety of their homes and communities. "Expert" witnesses such as corporate executives and government officials were not included, considering that we are inundated with their point of view and that those who have become experts through personal experience deserve equal time. For that very reason, some persons thoroughly conditioned by stereotypical representations of "experts" in broadcast media may view the documentary as not "objective" because those paid to represent corporate interests are not given the space to rebut the personal testimonials of citizen activists.

Why should they be afraid of us, we're ordinary citizens. I mean we're not eloquent. We know the facts, but I think that's what they're afraid of . . . that ordinary citizens are coming out and bringing up problems, highlighting them, raising up for everybody to see.[4]

An important objective of the project was to work with communities in such a way as to shift control of the discourse to those typically left out of the media. Through collaborative relationships in creative production, neighborhood political and cultural activists become active environmental communicators and technical producers, transforming their often eloquent, unheard voices and personal stories into political resistance.

Struggles for Voice and Power

Witness to the Future gives voice to citizens in communities where environmental disaster struck their back yards. In southeastern Washington State at the Hanford Nuclear Reservation, the plutonium used in the first atomic bomb was made in 1943. Hanford's releases of radioactive materials into air, water, and soil remained secret for more than 40 years. In 1986, because of public pressure by citizen activists, the U.S. Department of Energy released 19,000 classified documents proving that people were exposed to iodine-131 and other radioactive materials through contaminated food, air, water, and milk. Iodine-131 concentrates in the thyroid gland, where the most likely health effect is thyroid disease and cancer. Today, controversies continue over the cleanup of millions of gallons of highly radioactive waste still stored in tanks at Hanford.

In Louisiana's "Cancer Alley," stretching along the Mississippi River from Baton Rouge to New Orleans, over a hundred heavy industrial facilities provide a few jobs in the area, but they also release poison into the air, land, and water at a rate of almost half a billion pounds per year. The common sense of the local people, reflected in their calls for environmental justice, tells them that the petrochemical industry is about more than jobs; it is also brings sickness and death.

My ancestors and all the black people have been here in the United States of America since the year 1619. We have fought in every war this country ever been in, we have shed our blood and we're still not free. After these companies had been here about 15 years, that's when the dilemma started. . . . Every week we was marching behind two sometimes three people going to the graveyard in this little small community. If the people would just come out, every family that lost people in cancer and I guarantee you it would be declared as a cancer epidemic around here. And this is why they call it "Cancer Alley."[5]

Nearly half the produce consumed in the United States is grown in California's San Joaquin Valley, where farming is a $19-billion-a-year industry. Since their introduction in the 1950s, chemical pesticides have become standard in farming and have played an important role in the San Joaquin Valley's economy. Farmworkers and their families, who have direct, daily exposure to agricultural chemicals, suffer unusually high rates of cancer, birth defects, and environmental illnesses.[6]

This is what people who do not live near agriculture never think about. Every time you buy a piece of poison broccoli you are helping the farm worker get poisoned, you are helping some factory worker that works in some pesticide factory get poisoned, you are helping somebody downstream from that factory to get poisoned. There is a trail of poison from when the oil comes out of the ground, [is] turned into pesticides and ends up in . . . the food and then in ends up in our stomachs and causes all sort of problems from neurological disease to birth defects to cancer.[7]

The witnesses share stories of attempts to reach out to the institutions they expect will help them, only to be rejected, humiliated, disrespected, and lied to:

Not one penny has been spent on just trying to find people who were living here during those years, those decades and trying to alert them to their health risks. I mean they're not worried about downwinders—we're a dwindling, dying group. But they're worried about, obviously, credibility to the whole nuclear industry, credibility of the weapons-productions part of it and credibility of government itself is at issue here. So they don't want the public knowing about this stuff.[8]

Fighting illness and loss, and losing faith in the democratic ideals they were taught while growing up, their sense of abandonment, fear, and anger motivate them to take direct action. They realize that the only way to fight for the safety of their homes and community is to become their own best experts.

I don't pretend to be a scientist and I'm not an educated man, but I can read. And the more I read, and the more I understand, the more conferences that I attend, I realize that we were really used as human guinea pigs.[9]

As vocal, active role models participating in public hearings and gatherings, their fierce commitment challenges the authority of the "expert," for they are often not formally educated or in positions of power. In addition, by assuming many of the roles traditionally relegated in mediamaking to author and producer, they actively engage in transforming information into knowledge, and knowledge into action.

Subjects as Collaborators: Producing the Witness to the Future Video

As a model for media arts production, community building and education, and exploring new definitions of community participation, *Witness to the Future* brought together artists, nonprofit community centers, social service agencies, and environmental organizations. Attempting to transfer my expertise to those in the community, I gave up much of the control ordinarily maintained as a media maker.

As an outsider entering a community, I engage in a process of dialogue and exchange, creating trust and commitment. Although the final goal is a successful end result, collaboration with the community is of primary importance. This requires listening to them and establishing communication channels throughout the process that guarantee that the community takes charge of the message it chooses to deliver. The video was an experiment with multiple ways of transferring my role from autonomous "director" to that of facilitator, cultural producer, social organizer, and educator. A central challenge was to shift control to the subject while attempting to guarantee successful results, respecting their subjects' wisdom, strength, and ability to understand highly technical matters sufficiently well to make good technical and aesthetic decisions.

Script Development and Preproduction Strategies

The three sites were selected to deliver the message in a powerful way: so-called "ordinary" people can be transformed into environmental activists. In looking for communities representing diverse environmental crises, regions, economic strata, races, and cultures, the goal was to inspire the viewers to see themselves, their homes, and their neighborhoods in direct relation to the environmental struggles of the witnesses and communities selected.

The witnesses were almost always eager to share information, networking, and resources once we shared our goals. Early interviews with witnesses who had a high profile in their community led to the development of contacts who could "unveil" the stories of the area. It was also an opportunity to hear their voices, learn how they could contribute as subjects, assess their abilities to articulate their experiences in a dynamic, inspirational manner, and consider potential methods for collaboration.

Several witnesses did not speak English as their first language, and their interviews were conducted in their native tongue and subtitled in the video to respect their true voice. This diversity of language was a valuable contribution to the chorus of witnesses.

During the research, development, and preproduction process, local and regional environmental groups recommended citizen activists in the region. Production strategies were modeled to best achieve their goals and needs. After considerable attention to selecting individuals and communities, we agreed to collaborate, as they guided me in the development of the script and planning.

Production

Crews were often made up of nonprofessionals, students, community activists, and their friends. Familiar with the area, they helped access available resources. However, this also presented a challenge in learning to communicate with a new crew in each region. Each regional producer had a different way of communicating and operating, and each director of photography had a unique "eye," and work process.

To realize a collective vision, it is necessary to embrace the hybrid quality that comes out of such a grassroots production experience. This means merging low-tech and high-tech standards according to accessibility and remaining committed to the subject and ideas, rather than allowing them to be subservient to the technological demands of the medium. The freedom of not being limited by the hidden forms of media censorship, such as commercial sponsors or producers with financial affiliations, which profoundly influence a production, offers a unique and creative opportunity to articulate ideas that are critical and innovative.

Postproduction

In the editing phase the experimental documentary style was balanced with more conventional aspects of documentary form to reach the broadest audience possible. No voice-overs were used, so that the witnesses' voices framed their own representations. This decision resulted in a greater challenge in constructing a logical flow of information, but it grounded the work in the voice of the community.

The witnesses' stories were compelling, yet the nature of the medium required these stories (80 hours) to be condensed to fit the structure of the script outline and a bare-bones synthesis of the message. I wanted to let the material shape itself into an appropriate length, but had to weigh this against its intended use as a community organizing and education tool.

Interactive Tools for a New Generation: Audience as Cultural Producers of Information

After the documentary was completed, the obvious next step was to expand its content and use through interactive media, including a CD-ROM and the World Wide Web. This allowed us to include valuable information that had to be left out of the documentary.

The most exciting opportunity was that of expanding the collaboration from the subjects to the audience. The interactive form of the film transfers control of the content from the media maker to the viewer. In this new relationship, individuals alone and in collective viewing experiences can become aware of how the information was gathered and investigate its authenticity, breaking down, researching, and even reconstructing the information according to their own point of view.

As an organizing tool, an educational tool, and a tool for collective action, interactive media holds the potential to transform the viewer from a passive recipient to an active user and producer of information.

Viewers can obtain new "readings" in an active and participatory way. To gather and evaluate information, actively network with others, and even create their own environmental rhetoric.

Which tools would be truly valuable to citizen activists and community organizers, educators and students, as they work for environmental education and progressive change? A challenge was to encourage the CD-ROM user to go beyond the boundaries of pointing and clicking on the computer to select from the multiple choices so common in many interactive programs. The ability to link the CD-ROM to the Web allows the user to pursue his or her own research and define their own balance of objectivity through online research.

Critiquing the "Information Revolution's" myth that rapidly expanding amounts of information delivered through interactive technologies promise more informed audiences and empowered users, *Witness to the Future* is a contrast to new media technologies for military, industrial and entertainment sectors. Rather than focusing on the user as an individual, alone with their computer, the social end product is paramount. Instead of applying the "gaming" model relied upon in so many programs, encouraging competition and individual conquest, *Witness* stresses exploration and creativity, so that users become learners and sharers of knowledge. Interactive technologies conceived as tools for education and literacy can help viewers to think critically about information, transform that information into knowledge, and transfer that knowledge out of their computer and into their community.

An extraordinary expansion of the scope of *Witness to the Future* is the first electronic publication of Rachel Carson's *Silent Spring,* which is included in the CD-ROM. Originally published in 1962, Carson's work as a creative writer and scientist produced international debate on the subject of pesticides and effecting considerable change. During the production of the video, several of the witnesses noted that her masterpiece had served as their inspiration.

The reader of the electronic version can update Carson's prophetic visions with current events, and link to sites on the World Wide Web. Experiencing the same struggles as the witnesses, including attacks on her credibility, a long sickness, and ultimately death from breast cancer, we can hear her voice in an eloquent delivery of an exceptional 30-minute speech to the National Women's Press Club shortly after the publication of *Silent Spring* in 1962. Carson comments on many of the same issues the witnesses question, for example, how corporate power manipulates truth and interferes with our ability to learn the truth about environmental problems.

While investigating the real-life experiences of "ecoheroes" and "ecoheroines" who have struggled to clean up the environment in their homes, in their communities, and in the world, the viewer becomes an active producer and is encouraged to consider their own power to affect environmental change. By examining the successes, failures, and personal experiences of diverse communities, the multimedia producer can encourage the viewer to consider their own potential to work with others for environmental change. The user becomes their own best expert.

Screening and Distribution: The Social Context

To reach a grassroots audience, available resources are a crucial consideration: While interactive technologies are still in the hands of economically privileged, and even with access technical difficulties often interfere with the ability to interact effectively with an audience, consumer video (VHS) is available to a high percentage of the population in the United States and is easier to use. Better resolution resulting from screening in the video medium can result in a more intense experience for the audience. The linear delivery of presentation helps to guarantee an uninterrupted opportunity to present the issues.

Viewing such work in real, not only virtual space is vital. Public screenings, accompanied by live personal testimonials, create a network of cultural producers, community organizers, and concerned members of a region around related causes. In each region, rough-cut premieres followed by discussion brought together people with common concerns and shifted attention from the media itself to the active work in the community. In the premiere in New Orleans, the children from "Cancer Alley" presented live a scene from their script, adding to the dynamic exchange. At Junebug Production's Environmental Justice Festival of the Arts, which brought artists, activists, and community members together at the New Orleans Contemporary Arts Center, a witness in the final stages of lung cancer traveled to the screening to share personal struggles as an activist and mother, leaving not a dry eye in the place. These citizens' live presentations created a direct impact that would be impossible through the delivery of information alone. The shared experience produced a call for commitment that left the participants with a renewed need to stay active.

A premiere organized at Rensselaer Polytechnic Institute in Troy, New York, resulted in a grassroots public spectacle. To generate maximum publicity, we invited Lois Gibbs, the famous citizen activist and mother from Love Canal.[10] We also chose to use the premiere as a fund-raiser for a local citizens' environmental action group,[11] shifting attention and support to neighborhood environmental efforts. After contacting citizens' environmental groups from the surrounding region, a network of environmentally concerned organizations was developed which sent out an invitational letter and press release for the premiere. The event generated an audience of over 500 people, media coverage in every regional newspaper, and television coverage on all three local stations.

Whether applying the model of *Witness to the Future* in large or small groups, community or educational settings, conferences, workshops or festivals, it is essential to make connections between cultural production and community organizing. This includes building alliances with political and community organizers, attending to the impact that public awareness of the event can generate in the press and media, and considering the historical, cultural, economic, and environmental conditions of the community.

Cultural Production as Resistance: The Power of Participatory Practice

An active engagement in education stimulates the individual to consider their role as a member of a community and network of communities. It is through this process of reflection that the power of action can be unleashed, with citizens actively creating and disseminating their own environmental rhetoric, resulting in "cooperation, unity, organization and cultural synthesis."[12]

A collaborative design encourages an active process of "issue, investigation, and evaluation."[13] It gives recognition to the personal, including oral testimonials, aesthetic, and independent perspectives. This shifts attention from subject, constructed in the history of film and video practice through point of view, to the viewer's own point of view. Personal experience becomes a core function of the reception and interpretation of information.

We are calling out to all the communities: We fight for justice. We will not be moved. Do what your heart tells you. If you are prepared to fight, then go forward. We are aware that what we are doing, the battle we are fighting, we're not going to see results tomorrow or the next day or in a month or a year, but we are prepared not to be beaten.[14]

Creative practice is essential, not frivolous; cultural work is not the icing on the cake, it is a core aspect of organizing. The unnatural separation, which is often unquestioned, between artists and organizers, as well as binary constructs such as creative-analytical, diminishes the power of all of our actions. Human experience and emotion offer the cultural animation necessary to effect social change. Personal experience can inspire collective action. Everyone has a story to tell, or else we all remain silenced.

Notes

1. PAR is based on the existential concept of experience developed by Jose Ortega y Gasset, Spanish philosopher.
2. Paulo Friere, *Pedagogy of the Oppressed* (New York: Continuum, 1970, p. 108).
3. John Gaventa, "Toward a Knowledge Democracy: Viewpoints on Participatory Research in North America," in *Action and Knowledge: Breaking the Monopoly with Participatory Action-Research,* Orlando Falls-Borda, et al., eds. (New York: Apex Press, 1991, pp. 121–22).
4. Chris Gaudet, Pharmacists, St. Gabrielle, La, "Cancer Alley" section, *Witness to the Future.*
5. Amos Favorite, Sr., Ascension Parish Residents Against Toxic Pollution, Geismar, La, "Cancer Alley" section, *Witness to the Future.*
6. Supplemental sections, *Witness to the Future.*

7. George Raow, Volunteers for a Healthy Valley, Lompoc, California, San Joaquin Valley section, *Witness to the Future.*

8. Judith Jurgi, Hanford Downwinders Coalition, Seattle, Washington, Hanford Nuclear Reservation section, *Witness to the Future.*

9. Tom Bailie, farmer/downwinder, Mesa, Washington, Hanford Nuclear Reservation section, *Witness to the Future.*

10. Lois Gibbs is currently the executive director of the Center for Health, Environment and Justice in Virginia.

11. The Nassau Union of Concerned Citizens is a rural group located on the eastern border of New York State engaged in fighting hard-rock mining in their community.

12. Ibid., p. 8.

13. J. M. Ramsey, H. R. Hungerford, and T. Volk, "A Technique for Analyzing Environmental Issues," *Journal of Environmental Education* 21(1) (1989):26–30.

14. Community radio broadcast on KSJV, Radio Bilingue Fresno, "Parents for a Better Living/Padres Hacia Una Vida Mejor" (Buttonwillow, Calif., 1997).

Bibliography

Abram, Simon, Jacqueline Waldren, and Donald Macleod (eds.). *Tourists and Tourism: Identifying with People and Places.* Indianapolis: Berg Publishers, 1997.

Akwesasne Task Force on the Environment, R.A.C. "Superfund Clean-up at Akwesasne: A Case Study in Environmental Injustice," *International Journal of Contemporary Sociology* 34(2) (1997):267–290.

Alexander Ryan, Suzanne. "Companies Teach All Sorts of Lessons with Educational Tools They Give Away," *Wall Street Journal,* April 19, 1994.

Alperovitz, Gar et al. *Index of Environmental Trends: An Assessment of Twenty-One Key Environmental Indicators in Nine Industrialized Countries over the Past Two Decades.* Washington, D.C.: National Center for Economic Alternatives, 1995.

Alternatives for Community and Environment. *Beyond Vacant Lots Pilot Project: Dudley Square.* Boston: Alternatives for Community and Environment, July 1997.

Altvater, Elmar. *The Future of the Market.* New York: Verso, 1993.

Andersen, Robin. "Sport Utilities: Vehicles of Cultural Mythology," in *Critical Studies in Media Commercialism,* Robin Andersen and Lance Strate, eds. London: Oxford Univ. Press, 1999.

Andersen, Robin. *Consumer Culture and TV Programming.* Boulder, Col.: Westview, 1995.

Andrews, Cecile. *The Circle of Simplicity: Return to the Good Life.* New York: HarperCollins, 1997.

Arditti, Rita (ed.), *Science and Liberation.* Toronto: Univ. of Toronto Press, 1980.

Aronowitz, Stanley. *Science as Power: Discourse and Ideology in Modern Society.* Minneapolis: Univ. of Minnesota Press, 1988.

Ashford, Nicholas, and Claudia Miller. *Chemical Exposures: Low Levels and High Stakes.* New York: Van Nostrand Reinhold, 1990.

Athanasiou, Tom. *Divided Planet.* New York: Little, Brown, 1996.

Bachelard, Gaston. *The Poetics of Space.* Boston: Beacon Press, 1994.

Bacow, Lawrence S., and James R. Milkey. "Overcoming Local Opposition to Hazardous Waste Facilities," in *Resolving Locational Conflict,* Robert W. Lake, ed. New Brunswick, N.J.: Center for Urban Policy Research, 1987.

Bahro, Rudolf. *Avoiding Social and Ecological Disaster.* Bath, UK: Gateway Books, 1994, p. 50.

Bandy, Joe. "Managing the Other of Nature: Sustainability, Spectacle, and Global Regimes of Capital in Ecotourism," *Public Culture* 8(3) (1996):539–66.

Barker, Roger G. *Ecological Psychology: Concepts and Methods for Studying the Environment of Human Behavior.* Stanford, Calif.: Stanford Univ. Press, 1968.

Barnet, Richard J., and John Cavanagh. *Global Dreams: Imperial Corporations and the New World Order.* New York: Simon & Schuster, 1994.

Bartsch, Charles, and Richard Munson. "Restoring Contaminated Industrial Sites," *Issues in Science and Technology* (Spring 1994):74.

Bates, D.V. "The Effects of Air Pollution on Children," *Environmental Health Perspectives* **103**(6) (1995):49–53.

Beck, Ulrich. *The Risk Society: Towards a New Modernity.* London: Sage Publications, 1992.

Beder, Sharon. *Global Spin: The Corporate Assault on Environmentalism.* White River Junction, Vt: Green Books and Chelsea Green Publishing, 1998.

Berman, Daniel. *Death on the Job: Occupational Health and Safety Struggles in the United States.* New York: Monthly Review Press, 1978.

Blakeslee, Nate. "Carcinogenic Cornucopia," *Texas Observer,* January 30, 1998.

Bleifuss, Joel. "The Truth Hurts," *PR Watch* 3(4) (1996):1.

Brown, Michael E. *The Production of Society.* Totowa, N.J.: Rowman and Littlefield, 1986.

Bruno, Kenny. *Screening Foreign Investments: An Environmental Guide for Policy Makers and NGOs.* Penang, Malaysia: 1994.

Buchholz, Rogene A. *Principles of Environmental Management: The Greening of Business.* Englewood Cliffs, N.J.: Prentice-Hall, 1993.

Bullard, Robert D. *Unequal Protection.* San Francisco: Sierra Club Books, 1994.

Bullard, Robert D. "Anatomy of Environmental Racism and the Environmental Justice Movement," in *Confronting Environmental Racism: Voices from the Grassroots,* Robert Bullard, ed. Boston, Mass.: South End Press, 1993.

Bullard, Robert D. *Dumping in Dixie: Race, Class, and Environmental Quality,* 2d ed. Boulder, Col.: Westview Press, 1994.

Camacho, David. *Environmental Injustices, Political Struggles: Race, Class and the Environment.* Durham, N.C.: Duke Univ. Press, 1998.

Canguilhem, Georges. *The Normal and the Pathological.* New York: Zone Books, 1991.

Carey, Alex. *Taking the Risk Out of Democracy: Propaganda in the US and Australia.* Sydney, Australia: Univ. of New South Wales Press, 1995.

Carley, Michael, and Philippe Spapens. *Shaping the World: Sustainable Living and Global Equity in the 21st Century.* London: Earthscan, 1998.

Carson, Rachel. *Silent Spring.* Boston: Houghton Mifflin, 1962.

Centers for Disease Control and Prevention. "Surveillance for Asthma—United States, 1960–95." *Morbidity and Mortality Weekly Report* 47, SS-1 (1998).

Chatterjee, Pratap, and Matthias Finger. *The Earth Brokers: Power, Politics, and World Development.* New York: Routledge, 1994.

Chivian, Eric et al. *Critical Condition: Human Health and the Environment.* Cambridge, Mass.: MIT Press, 1994.

Cohen, Gary, and John O'Connor. *Fighting Toxics.* Washington, D.C.: Island Press, 1990.

Colborn, Theo et al. *Our Stolen Future: Are We Threatening Our Fertility, Intelligence, and Survival?—A Scientific Detective Story.* New York: Dutton, 1996.

Colditz, G. A. "Family History, Age, and Risk of Breast Cancer: Prospective Data from the Nurses' Health Study," *Journal of the American Medical Association* 2(70) (1993):338–43.

Commoner, Barry. *Making Peace with the Planet.* New York: New Press, 1992.

Coyle, Marcia, and Marianne Lavelle. "Unequal Protection: The Racial Divide in Environmental Law," *National Law Journal* (September 21, 1992).

Cranor, Carl F. *Regulating Toxic Substances: A Philosophy of Science and the Law.* Oxford: Oxford Univ. Press, 1993.

Crossen, Cynthia. *Tainted Truth: The Manipulation of Fact in America.* New York: Simon & Schuster, 1994.

Curtis, S. A. "Cultural Relativism and Risk-Assessment Strategies for Federal Projects," *Human Organization* 51(1) (1992):65–70.

Deetz, Stanley. *Democracy in the Age of Corporate Colonization: Developments in Communication and the Politics of Everyday Life.* Albany, N.Y.: State Univ. of New York Press, 1992, p. 88.

DeLuca, Kevin. "Constituting Nature Anew through Judgment: The Possibilities of Media," in *Earthtalk: Communication Empowerment for Environmental Action,* Star A. Muir and Thomas Veenendall, eds. Westport, Conn. Praeger, 1996.

Dembe, Allard E. *Occupation and Disease: How Social Factors Affect the Conception of Work Related Disorders.* New Haven, Conn.: Yale Univ. Press, 1996.

Dowie, Mark. *Losing Ground: American Environmentalism at the Close of the 20th Century.* Cambridge, Mass.: MIT Press, 1995.

Doyle, Jack. *Hold the Applause!: A Case Study of Corporate Environmentalism as Practiced at DuPont.* Washington, D.C.: Friends of the Earth, 1991.

Durning, Alan. *How Much Is Enough? The Consumer Society and the Future of the Earth.* New York: W. W. Norton, 1992.

Ekins, Paul. "The Sustainable Consumer Society: A Contradiction in Terms?" *International Environmental Affairs* 3(4) (Fall 1991):243–58.

Elling, Ray. *The Struggle for Worker Health: A Study of Six Industrialized Countries.* New York: Baywood, 1986.

Engels, Frederick. *The Condition of the Working Class in England* (1848). Stanford, Calif.: Stanford Univ. Press, 1958.

Environmental Protection Agency. *1992 Toxic Chemical Release Inventory: Public Data Release.* EPA 745-R-001, Washington, D.C.: EPA, 1994, p. 79.

Epstein, Samuel S. "Environmental and Occupational Pollutants Are Avoidable Causes of Breast Cancer," *International Journal of Health Services* 24 (1994): 145–50.

Faber, Daniel (ed.). *The Struggle for Ecological Democracy: Environmental Justice Movements in the United States.* New York: Guilford Press, 1998.

Ferris, Deeohn. "Future Use Would Continue Past Inequities," *Environmental Forum* 10(6) (November/December 1993):34ff.

Fish, Stanley. "How the Right Hijacked the Magic Words," Op-Ed, *New York Times,* August 13, 1995, p. E15.

Fitzgerald, E., S-A. Hwang, K. Brix, B. Bush, K. Cook, and P. Worswick. "Fish PCB Concentrations and Consumption Patterns Among Mohawk Women at Akwesasne," *Journal of Exposure Analysis and Environmental Epidemiology* 5(1) (1995):1–19.

Foderano, Lisa W. "For Affluent Blacks, Harlem's Pull Is Strong," *New York Times,* January 18, 1998, pp. A1, A25.

Foster, John Bellamy. "The Limits of Environmentalism without Class: Lessons from the Ancient Forest Struggle in the Pacific Northwest," in *The Struggle for Ecological Democracy,* Daniel Faber, ed. New York: Guilford Press, 1998, pp. 188–217.

Fried, John J. "Firms Gain Hold in Environmental Education," *Philadelphia Inquirer,* March 27, 1994.

Friere, Paulo. *Pedagogy of the Oppressed.* New York: Continuum, 1970.

Fullilove, Robert E., Mindy T. Fullilove, Mary E. Northridge, Michael L. Ganz, Mary T. Bassett, Diane E. McLean et al. "Risk Factors for Excess Mortality in Harlem: Findings from the Harlem Household Study," *American Journal of Preventive Medicine* 16(3S) (1999):22–28.

Fullilove, Mindy Thompson. "Psychiatric Implications of Displacement: Contributions from the Psychology of Place," *American Journal of Psychiatry* 153 (1996):1516–23.

Fullilove, Mindy Thompson. "Environments as Toxins: Comment on Panel on Strengths and Potentials of Adolescence," *Bulletin of the New York Academy of Medicine* 67 (1991):571–73.

Gallagher, Winifred. *The Power of Place.* New York: HarperPerennial, 1994.

Gare, Arran E. *Postmodernism and the Environmental Crisis.* New York: Routledge, 1995.

Gaventa, John. "Toward a Knowledge Democracy: Viewpoints on Participatory Research in North America," in *Action and Knowledge: Breaking the Monopoly with Participatory Action-Research,* Orlando Falls-Borda, et al., eds. New York: Apex Press, 1991.

Gedicks, Al. *The New Resource Wars: Native and Environmental Struggles Against Multinational Corporations.* Boston: South End Press, 1993.

Geiser, Ken. "Protecting Reproductive Health and the Environment: Toxics Use Reduction," *Environmental Health Perspectives* 101 (Suppl. 2) (1993):221–25.

Geiser, Ken. "The Greening of Industry: Making the Transition to a Sustainable Economy," *Technology Review* (August/September 1991:65–72.

Geras, Norman. "Essence and Appearance: Aspects of Fetishism in Marx's Capital," in *Ideology in Social Science,* Robin Blackburn, ed. London: Penguin, 1972.

Gerrard, Michael B. *Whose Backyard, Whose Risk: Fear and Fairness in Toxic and Nuclear Waste Siting.* Cambridge, Mass.: MIT Press, 1994.

Goldman, Benjamin. *The Truth About Where You Live: An Atlas for Action on Toxins and Mortality.* New York: Random House, 1992.

Goldman, Robert, and Stephen Papson. *Sign Wars: The Cluttered Landscape of Advertising.* New York: Guilford Press, 1996.

Goldtooth, Tom. "Indigenous Nations: Summary of Sovereignty and Its Implications for Environmental Protection," in *Environmental Justice: Issues, Policies and Solutions,* Bunyan Bryant, ed. Washington, D.C.: Island Press, 1995.

Gottlieb, Robert. *Reducing Toxics: A New Approach to Policy Making and Industrial Decisionmaking.* Washington, D.C.: Island Press, 1995.

Gramsci, Antonio. *Selections from the Prison Notebooks.* New York: International Publishers, 1971.

Grefe, Edward A., and Martin Linsky. *The New Corporate Activism.* New York: McGraw-Hill, 1995.

Greider, William. *Who Will Tell the People.* New York: Simon & Schuster, 1992.

Haineault, Doris-Louis, and Jean-Yves Roy. *Unconscious for Sale: Advertising, Psychoanalysis and the Public.* Minneapolis: Univ. of Minnesota Press, 1993.

Hall, R. R. "Superficial Bladder Cancer," *British Medical Journal* 308 (1994): 910–13.

Hall, Stuart. "The Problem of Ideology: Marxism Without Guarantees," in *Stuart Hall: Critical Dialogues in Cultural Studies,* David Morley and Kuan-Hsing Chen, eds. New York: Routledge, 1996.

Harding, Sandra. *Whose Science, Whose Knowledge?* Ithaca, N.Y.: Cornell Univ. Press, 1991.

Harper, B. L. "Incorporating Tribal Cultural Interests and Treaty-Reserved Rights in Risk Management," in *Fundamentals of Risk Analysis and Risk Management*, V. Molak, ed. Boca Raton, Fla.: Lewis Publishers, 1997.

Harrison, E. Bruce. "Managing for Better Green Reputations," *International PR Review* **17**(3) (1994):8.

Hauptman, Lawrence. "Drums Along the Waterways: The Mohawks and the Coming of the St. Lawrence Seaway," in *The Iroquois Struggle for Survival*, Lawrence Hauptman, ed. Syracuse, N.Y.: Syracuse Univ. Press, 1986.

Heller, Chaia. "For the Love of Nature: Ecology and the Cult of the Romantic," in *Ecofeminism: Women, Animals and Nature*, Greta Gaard, ed. Philadelphia: Temple Univ. Press, 1993.

Jacobs, Michael. "The Limits to Neoclassicism: Towards an Institutional Environmental Economics," in *Social Theory and the Environment*, Michael Redclift and Ted Benton, eds. New York: Routledge, 1994.

Jameson, Frederic. *The Political Unconscious: Narrative as a Socially Symbolic Act*. Ithaca, N.Y.: Cornell Univ. Press, 1981.

Jameson, Frederic. *Postmodernism or the Cultural Logic of Late Capitalism*. Durham, N.C.: Duke Univ. Press, 1991.

Johnson, J. "Mohawk Environmental Health Project Integrates Research into the Community," *Environmental Science and Technology* **30**(1) (1996):20A.

Kane, M. J. "Promoting Political Rights to Protect the Environment, "*Yale Journal of International Law* **18** (1993):389–411.

Kapp, K. William. *The Social Costs of Private Enterprise*. New York: Schocken Books, 1971, p. 231.

Kettl, P. "Suicide and Homicide: The Other Costs of Development," *Northeast Indian Quarterly* **84** (1991):58.

King, David, and William Stewart. "Ecotourism and Commodification: Protecting People and Places," *Biodiversity and Conservation*, **5**(3) (1996):293–305.

Korten, David. *When Corporations Rule the World*. West Hartford, Conn.: Kumarian Press, 1995.

Kothari, Rajni. *Growing Amnesia: An Essay on Poverty and the Human Consciousness*. New Delhi: Viking Penguin India, 1993.

Krieger, Nancy. "Exposure, Susceptibility, and Breast Cancer Risk," *Breast Cancer Research and Treatment* **13** (1989):205–23.

Krook, L., and G. A. Maylin. "Industrial Fluoride Pollution: Chronic Fluoride Poisoning in Cornwall Island Cattle," *Cornell Veterinarian* **69** (Suppl. 8) (1979): 1–70.

Kuhn, Sarah, and John Wooding. "The Changing Structure of Work in the U.S.: Part 1—The Impact on Income and Benefits," *New Solutions: A Journal of Environmental and Occupational Health Policy* **4**(2) (1994):43–56.

Kuhn, Sarah, and John Wooding. "The Changing Structure of Work in the U.S.: Part II—The Implications for Health and Welfare," *New Solutions: A Journal of Environmental and Occupational Health Policy* 4(4) (1994):21–27.

Kuntz, Tom. "The McLibel Trial," *New York Times,* August 6, 1995, p. E7.

Lappé, Marc. *Chemical Deception: The Toxic Threat to Health and the Environment.* San Francisco: Sierra Club Books, 1991.

Lax, Michael. "Workers and Occupational Safety and Health Professionals: Developing the Relationship," *New Solutions: A Journal of Environmental and Occupational Health Policy* 8(1) (1998):99–116.

Lee, Martin J. *Consumer Culture Reborn: The Cultural Politics of Consumption.* New York: Routledge, 1993.

Lefebvre, Henri. *Everyday Life in the Modern World.* New York: Harper & Row, 1971.

Legault, Leonard et al. Ninth Biennial Report on Great Lakes Water Quality. Washington, D.C.: International Joint Commission, 1998.

Levenstein, Charles, John Wooding, and Beth Rosenberg. "The Social Context of Occupational Health," in *Occupational Health: Recognizing and Preventing Work-Related Diseases,* Barry S. Levy and David H. Wegman, eds. Boston: Little, Brown, 1995.

Levy, Barry S., and David H. Wegman (eds.). *Occupational Health: Recognizing and Preventing Work-Related Diseases.* Boston: Little, Brown, 1995.

Luke, Timothy. *Ecocritique: Contesting the Politics of Nature, Economy, and Culture.* Minneapolis: Univ. of Minnesota Press, 1997.

Luke, Timothy. "Green Consumerism: Ecology and the Ruse of Recycling," in *In the Nature of Things: Language, Politics and the Environment,* Jane Bennett and William Chaloupka, eds. Minneapolis: Univ. of Minnesota Press, 1993.

MacCannell, Dean. *The Tourist: A New Theory of the Leisure Class.* New York: Schocken Books, 1989.

Mailander, Jodi, and Cyril T. Zanesky. "Big Business and the Classroom," *Miami Herald,* April 17, 1994, p. 1A.

Mander, Jerry. *In the Absence of the Sacred: The Failure of Technology and the Survival of the Indian Nations.* San Francisco: Sierra Club Books, 1991.

Mann, Eric. *L.A.'s Lethal Air.* Los Angeles: Labor/Community Strategy Center, 1991.

Mann, Jonathan, Daniel J. M. Tarantola, and Thomas W. Netter (eds.). *AIDS in the World: A Global Report.* Cambridge, Mass.: Harvard Univ. Press, 1992.

Markowitz, Gerald, and David Rosner. "Pollute the Poor," *The Nation,* July 6, 1998, p. 8.

Mazmanian, Daniel, and David Morell. *Beyond Superfailure: America's Toxics Policy for the 1990's.* Boulder, Col.: Westview, 1992.

May, Peter H., and Ronaldo Seroa Motta (eds.). *Pricing the Planet.* New York: Columbia Univ. Press, 1996.

McAllister, Matthew. *The Commercialization of American Culture: New Advertising, Control and Democracy.* Thousand Oaks, Ca.: Sage Publications, 1996.

McChesney, Robert W., Ellen Meiksins Wood, and John Bellamy Foster (eds.). *Capitalism and the Information Age: The Political Economy of the Global Communication Revolution.* New York: Monthly Review Press, 1998.

Mclaren, Deborah. *Rethinking Tourism and Ecotravel.* West Hartford, Conn.: Kumarian Press, 1998.

McLuhan, Marshall. *The Mechanical Bride: Folklore of Industrial Man.* New York: Basic Books, 1967.

Merrell, P., and C. Van Strum. "Negligible Risk: Premeditated Murder?" *Journal of Pesticide Reform* 10 (1990):20–22.

Messaris, Paul. *Visual Persuasion: The Role of Images in Advertising.* Thousand Oaks, Ca.: Sage Publications, 1997.

Mollenkopf, John. *The Contested City.* Princeton, N.J.: Princeton Univ. Press, 1983.

Montague, Peter. "The Many Uses of Risk Assessment," *Rachel's Environment & Health Weekly* No. 420 (December 15, 1994).

Montague, Peter. "The Breakdown of Morality," *Rachel's Hazardous Waste News* No. 287. (May 27, 1992).

Mort, Frank. "The Politics of Consumption," in *New Times: The Changing Face of Politics in the 1990s,* Stuart Hall and Martin Jacques, eds. New York: Verso, 1990.

Munt, Ian. "Eco-tourism or Ego-tourism?" *Race & Class* 36(1) (1994):49–60.

National Environmental Justice Advisory Council, Federal Advisory Committee to the U.S. Environmental Protection Agency, *Environmental Justice, Urban Revitalization, and Brownfields: The Search for Authentic Signs of Hope,* Charles Lee, ed. EPA 500-R-96-002. December 1996.

Navarro, Vincente. *Crisis, Health, and Medicine: A Social Critique.* New York: Tavistock Publications, 1986.

Noble, Charles. *Liberalism at Work: The Rise and Fall of OSHA.* Philadelphia: Temple Univ. Press, 1986.

Nordhaus, William. "Expert Opinion on Climate Change," *American Scientist* 82(1) (January/February 1994):4–51.

O'Brien, Mary. "Alternatives to Risk Assessment: The Example of Dioxin," *New Solutions* (Winter 1993):39–42.

O'Brien, Mary. "A Crucial Matter of Cumulative Impacts: Toxicity Equivalency Factors," *Journal of Pesticide Reform* 10(2) (1990):23–27.

O'Connor, Martin. "On the Misadventures of Capitalist Nature," in *Is Capitalism Sustainable?: Political Economy and the Politics of Ecology*, Martin O'Connor, ed. New York: Guilford Press, 1994, pp. 125–151.

O'Riordan, T., and J. Cameron (eds.). *Interpreting the Precautionary Principle*. London: Earthscan, 1994.

Orlow, I. et al. "Deletion of the p16 and p15 Genes for Human Bladder Tumors," *Journal of the National Cancer Institute* 87 (1995):1524–29.

Ouellet-Hellstromt, R., and J. D. Rench, "Bladder Cancer Incidence in Arylamine Workers," *Journal of Occupational and Environmental Medicine* 38 (1996): 1239–47.

Panos Institute, "Green or Mean?: Environment and Industry Five Years on from the Earth Summit," Panos Media Briefing No. 24. London: Panos, June 1997.

Pearce, D. W. *Environmental Economics*. New York: Longman, 1974.

Pearson, Charles S. (ed.). *Multinational Corporations, Environment and the Third World*. Durham, N.C.: World Resources Institute/Duke Univ. Press, 1987.

Perera, R. "Uncovering New Clues to Cancer Risk," *Scientific American* (May 1996):54–62.

Philips, Peter (ed.), *Censored 1998: The News that Didn't Make the News*. New York: Seven Stories Press, 1998.

Pimentel, David. "Ecology of Increasing Disease," *BioScience* 48(1) (October 1998); 817.

Pirkle, J. L. et al. "The Decline of Blood Lead Levels in the United States: The National Health and Nutrition Examination Surveys," *Journal of the American Medical Association* 272 (1994):284–91.

Polayni, Karl. *The Great Transformation*. Boston: Beacon Press, 1944.

Powers, C., and M. Chertow, "Industrial Ecology," in *Thinking Ecologically: The Next Generation of Environmental Policy*, M. Chertow and D. Esty, eds. New Haven, Conn.: Yale Univ. Press, 1997.

Prager, Sharon. "Changing North America's Mind-Set about Mining," *Engineering and Mining Journal* 198(2) (February 1997):36–44.

Proctor, R. N. *Cancer Wars: How Politics Shapes What We Know and Don't Know about Cancer*. New York: Basic Books, 1995.

"Public Interest Pretenders," *Consumer Reports*, May 1994, p. 316.

Pulido, Laura. "Ecological Legitimacy and Cultural Essentialism: Hispano Grazing in the Southwest," in *The Struggle for Ecological Democracy: Environmental Justice Movements in the United States*, Daniel Faber, ed. New York: Guilford Press, 1998.

Pulido, Laura. *Environmentalism and Economic Justice: Two Chicano Struggles in the Southwest*. Tucson, Ariz.: Univ. of Arizona Press, 1996.

"Philadelphia Dumps on the Poor," *Rachel's Environment and Health Weekly* No. 595 (April 23, 1998).

Ramsey, J. M., H. R. Hungerford, and T. Volk. "A Technique for Analyzing Environmental Issues," *Journal of Environmental Education* 21(1) (1989):26–30.

Rench, J. D. et al. *Cancer Incidence Study of Workers Handling Mono- and Diarylamines Including Dichlorobenzidine, ortho-Toluidine, and ortho-Dianisidine.* Falls Church, Va.: SRA Technologies, 1995.

Reynolds, Richard. "Mother Jones Releases USDA Memo Detailing Plans to Gut NOSB Recommendations on Organic Standards," press release. San Francisco: *Mother Jones Magazine* (March 12, 1998).

Roe, David, William Pease, Karen Florini, and Ellen Silbergeld. *Toxic Ignorance: The Continuing Absence of Basic Health Testing for Top-Selling Chemicals in the United States.* New York: Environmental Defense Fund, 1997.

Roht-Arriaza, Naomi. UNCED Undermined: Why Free Trade Won't Save the Planet. Greenpeace UNCED Reports, Greenpeace International (March 1992).

Rose, Merrill. "Activism in the 90s: Changing Roles for Public Relations," *Public Relations Quarterly* 36(3) (1991):28–32.

Rosner, David, and Gerald Markowitz (eds.). *Dying for Work.* Bloomington: Indiana Univ. Press, 1987.

Rowell, Andrew. *Green Backlash: Global Subversion of the Environmental Movement.* London: Routledge, 1996.

Sachs, Wolfgang. "The Blue Planet: An Ambiguous Modern Icon," *Ecologist* 24(5) (September/October 1994):170–75.

Sachs, Wolfgang, Reinhard Loske, Manfred Linz et al. *Greening the North: A Post-Industrial Blueprint for Ecology and Equity.* New York: Zed Books, 1998.

Sagoff, Mark. *The Economy of the Earth.* New York: Cambridge Univ. Press, 1988.

Sanders, Barry. *The Private Death of Public Discourse.* Boston: Beacon Press, 1998.

Sassen, Saskia. *Globalization and Its Discontents.* New York: New Press, 1998.

Schiller, Herbert I. *Culture Inc.: The Corporate Takeover of Public Expression.* New York: Oxford Univ. Press, 1989.

Schor, Juliet B. *The Overspent American: Upscaling, Downshifting, and the New Consumer.* New York: Basic Books, 1998.

Schulte, R. A. et al. (eds.). "Bladder Cancer Screening in High Risk Groups," *Journal of Occupational Medicine* 32 (1990):787–945.

Schumacher, E. F. *Small Is Beautiful.* New York: Harper & Row, 1973.

Shiva, Vandana. "The Greening of the Global Reach," *Third World Resurgence* No. 14/15 (1992) (Penang, Malaysia).

Shkilnyk, A. M. *A Poison Stronger than Love: The Destruction of an Ojibwa Community.* New Haven: Yale Univ. Press, 1985.

Shoshone-Bannock Tribes. "Risk Assessment in Indian Country: Guiding Principles and Environmental Ethics of Indigenous Peoples," Position paper written as part of the National Tribal Risk Assessment Forum, "An Indigenous Approach to Decision Making," 1996.

Simon, Stephanie. "Exploring California: Tourism with a Message," *Los Angeles Times,* February 15, 1998: Home Edition, Part A, Column 1.

Smith, Barbara Ellen. "Black Lung: The Social Production of Disease," *International Journal of Health Services* 11 (1981):343-59.

Smith, Zachary A. *The Environmental Policy Paradox,* 2d ed. Englewood Cliffs, N.J.: Prentice-Hall, 1995.

Spears, Ellen Griffith. *The Newtown Story: One Community's Fight for Environmental Justice.* Gainesville, Ga.: The Center for Democratic Renewal and the Newtown Florist Club, 1998.

Stauber, John, and Sheldon Rampton. *Toxic Sludge Is Good for You.* Monroe, Maine: Common Courage Press, 1995.

Steingraber, Sandra. *Living Downstream: An Ecologist Looks at Cancer and the Environment.* Reading, Mass.: Addison-Wesley, 1997.

Stillwagon, Eileen. *Stunted Lives, Stagnant Economies: Poverty, Disease, and Underdevelopment.* New Brunswick, N.J.: Rutgers Univ. Press, 1998.

Stisser, Peter. "A Deeper Shade of Green," *American Demographics* 16(3) (March 1, 1994):24.

Stone, Ward. "New York State Assembly Hearings: Crisis at Akwesasne, Day II, Transcript," in *Ecocide of Native America.* D. Grinde and B. Johansen, eds. Santa Fe, N.M.: Clear Light Publishers, 1995.

Sunday, J. *The Community of Akwesasne.* Mohawk Territory of Akwesasne. Department of Environment, Mohawk Council of Akwesasne, 1996.

Swartz, David. *Culture and Power: The Sociology of Pierre Bourdieu.* Chicago: Univ. of Chicago Press, 1997.

Szasz, Andrew. *Ecopopulism: Toxic Waste and the Movement for Environmental Justice.* Minneapolis: Univ. of Minnesota Press, 1994.

Tarlau, Eileen Senn. "Playing Industrial Hygiene to Win," *New Solutions: A Journal of Environmental and Occupational Health Policy* 1(4) (Spring 1991):72–81.

Tarlov, Alan R. "Social Determinants of Health: The Sociobiological Translation," in *Health and Social Organization: Towards a Health Policy for the 21st Century,* David Blane, Eric Brunner, and Richard Wilkinson, eds. New York: Routledge, 1996, pp. 71–93.

Taylor, Bron Raymond (ed.). *Ecological Resistance Movements: The Global Emergence of Radical and Popular Environmentalism.* Albany, N.Y.: State Univ. of New York Press, 1995.

Tokar, Brian. *Earth for Sale: Reclaiming Ecology in the Age of Corporate Greenwash*. Boston: South End Press, 1997.

Tomlinson, John. *Cultural Imperialism*. Baltimore, Md.: Johns Hopkins Univ. Press, 1991, p. 168.

Tomosho, R. "Dumping Grounds: Indian Tribes Contend with Some of Worst of America's Pollution," *Wall Street Journal*, November 29, 1990, pp. A1, A6.

Torres, Gerald. "Environmental Burdens and Democratic Justice," *Fordham Urban Law Journal* **21**(3) (Spring 1994):431–460.

Trichopoulos, D. et al. "What Causes Cancer?" *Scientific American* (September 1996):80–87.

Turner, R. Kerry, David Pearce, and Ian Bateman. *Environmental Economics*. Baltimore, Md.: Johns Hopkins Univ. Press, 1993.

Turshen, Meredeth. *The Politics of Public Health*. New Brunswick, N.J.: Rutgers Univ. Press, 1989.

Wallace, Roderick. "A Synergism of Plagues: 'Planned Shrinkage,' Contagious Housing Destruction, and AIDS in the Bronx," *Environmental Research* **47** (1988):1–33.

Walsh, Alison. "Forever Fresh: Lost in the Land of Feminine Hygiene," *Utne Reader* (September/October 1996):32.

Ward, Milton. "Mining and the Environment: In the Long Run We Are All . . . Survivors," *Bulletin of the Institution of Mining and Metallurgy* No. 1006 (May 1992):33, 35.

Waring, Marilyn. *If Women Counted*. New York: HarperCollins, 1988.

Wilkinson, Richard J. *Unhealthy Societies: The Affliction of Inequality*. New York: Routledge, 1996.

Williams, Raymond. *Marxism and Literature*. London: Oxford Univ. Press, 1977.

Willums, Jan-Olaf, and Ulrich Goluke. *From Ideas to Action: Business and Sustainable Development, The Greening of Enterprise 1992*. International Environmental Bureau of the International Chamber of Commerce, Norway, May 1992.

Wooding, John, and Charles Levenstein (eds.). *Work, Health and Environment: Old Problems, New Solutions*. New York: Guilford Press, 1998.

World Resources Institute. *Corporate Environmental Accounting*. Washington, D.C.: WRI, 1995.

Young, John E. "Mining the Earth," Worldwatch Paper No. 109. Washington, D.C.: Worldwatch Institute, 1992.

Contributors

Robin Andersen is associate professor, Department of Communication and Media Studies, and director of the Peace and Justice Studies Program at Fordham University. She is the author of *Consumer Culture and TV Programming* and has published numerous articles in books and scholarly journals.

Mary Frances Arquette works for the St. Regis Mohawk Tribe's environment division as the coordinator for the Natural Resource Damage Program. Dr. Arquette received a Ph.D. in environmental toxicology from Cornell University in 1994 and a doctor of veterinary medicine from the New York State College of Veterinary Medicine in 1990. She has worked as an assistant professor of environmental studies and has published articles on Native American environmental issues as well as canine immunotoxicology.

Marcy Darnovsky is an adjunct faculty member at the Hutchins School of Liberal Studies at Sonoma State University (California). Her academic work has been published in *Socialist Review* and *Social Text* and in the anthology *Cultural Politics and Social Movements,* which she co-edited with Barbara Epstein and Dick Flacks.

Giovanna Di Chiro is assistant professor of environmental studies at Allegheny College in Meadville, Pennsylvania. Her work has focused on environmental and social justice issues from cross-cultural and international perspectives.

John Bellamy Foster is professor of sociology, University of Oregon, Eugene. He is an editor of the journal *Capitalism, Nature, Socialism,* and the author of a number of books, including *The Vulnerable Planet* and *The Theory of Monopoly Capitalism.* Dr. Foster is also co-editor of *The Faltering Economy.*

Robert E. Fullilove III is the associate dean for community and minority affairs and associate professor of clinical public health in sociomedical sciences at the Joseph A. Mailman School of Public Health of Columbia University. He currently directs the master's program in health promotion and disease prevention in the Division of Sociomedical Sciences at Columbia and is co-director of the Community Research Group. In 1998 he was named a visiting Falk Fellow (along with his wife, Mindy Thompson Fullilove, M.D.) at the University of Pittsburgh Graduate School of Public Health.

Al Gedicks   is a professor of sociology at the University of Wisconsin, LaCrosse. He is the author of *The New Resource Wars: Native and Environmental Struggles Against Multinational Corporations.*

Richard Hofrichter   is a writer and social critic. He is the author of *Neighborhood Justice in Capitalist Society* and the editor of *Toxic Struggles: The Theory and Practice of Environmental Justice.* He received a Ph.D. in political science from the City University of New York in 1983.

Joshua Karliner   is founder and executive director of the Transnational Action and Resource Center (TRAC) in San Francisco. He also serves as editorial coordinator of the Corporate Watch Web site. He is the author of the *Corporate Planet: Ecology and Politics in the Age of Globalization.* Mr. Karliner has taught environmental politics at the University of San Francisco and served as Earth Summit coordinator for Greenpeace International.

Charles Levenstein   is professor of work environment policy at the University of Massachusetts, Lowell. Dr. Levenstein is editor of *New Solutions,* an international journal of occupational and environmental health policy. His latest book, co-authored with John Wooding, is *The Point of Production: The Political Economy of the Work Environment.* He has another forthcoming book co-edited with John Wooding, *New Solutions: A Reader in Work Environment Policy.* He is principal investigator of a project funded by the National Institute for Environmental Health Sciences that provides health and safety training to hazardous waste workers and emergency responders.

Timothy W. Luke   is professor of political science at Virginia Polytechnic Institute in Blacksburg, Virginia. He is the author of *Ecocritique: Contesting Nature, Economy, and Culture,* and *Capitalism, Democracy, and Ecology: Departing from Marx.*

Rafael Mares   a graduate of Harvard Law School, is a Sacks Fellow in the housing unit of Harvard Law School's Legal Services Center in Jamaica Plain, Massachusetts, where he represents low-income tenants who suffer from asthma and lead poisoning. He is also co-author of *Environmental Discrimination,* a Council of Planning Librarians annotated bibliography, and a founder of the first environmental justice course at the University of Vermont.

Branda Miller   is associate professor of video at Rensselaer Polytechnic Institute, Troy, New York. She is a video artist, activist, and media educator who has been working with independent forms of media and electronic arts for over 15 years. Her most recent work is *Witness to the Future.*

Mary H. O'Brien   is chair of the toxics board of the City of Eugene, Oregon. She is also a consultant on alternatives to risk assessment and reduction of toxics use and is ecosystem policy analyst for the Hells Canyon Preservation Council, Joseph, Oregon. She has published dozens of articles in books and journals, and her new book, *Making Better Decisions: An Alternative to Risk Assessment,* is published by the MIT Press.

John O'Neal is an award-winning playwright, actor, director, and teacher. He is artistic director of Junebug Productions, a performing arts organization based in New Orleans. Mr. O'Neal was a co-founder of the Free Southern Theater, whose philosophy was "to use theater as an instrument to stimulate the development of critical and reflective thought among Black people in the South." Mr. O'Neal has collaborated in the writing and performance of a highly acclaimed trilogy known collectively as *Sayings from the Life and Writings of Junebug Jabbo Jones*. He has recently completed a three-year teaching position as visiting professor in theater at Cornell University.

Sheldon Rampton is associate editor of the *PR Watch* Web site in Madison, Wisconsin. He is the co-author (with John Stauber) of *Toxic Sludge Is Good for You* and *Mad Cow USA: Could the Nightmare Happen Here?* He is also co-author of *Friends in Need: The Story of U.S. Nicaragua Sister Cities* and *Trust Us, We're Experts,* forthcoming.

William Shutkin is president of New Ecology, Inc., a catalyst for environmental protection, economic development, and community-building strategies in New England based in Cambridge, Massachusetts. Mr. Shutkin is also an adjunct assistant professor of law at Boston College Law School, where he has taught environmental law since 1994, and a lecturer in the Urban and Environmental Policy Department at Tufts University. His new book, *The Land That Could Be: Environmentalism and Democracy in the Twenty-First Century,* is published by the MIT Press.

John Stauber is an author, public speaker, social activist, and the founder and executive director of the Center for Media and Democracy in Madison, Wisconsin. He is the co-author (with Sheldon Rampton) of *Toxic Sludge Is Good for You* and *Mad Cow USA: Could the Nightmare Happen Here?* He is also co-author of *Trust Us, We're Experts,* forthcoming.

Sandra Steingraber is a biologist and cancer survivor. Currently a visiting professor at Cornell University's Program for Breast Cancer and Environmental Risk Factors, she is the author of *Living Downstream: An Ecologist Looks at Cancer and the Environment,* from which her chapter is adapted. She also serves on the National Action Plan on Breast Cancer, administered by the U.S. Department of Health and Human Services.

Alice Tarbell is the director of the First Environmental Research Projects of the Akwesasne Task Force on the Environment, an environmental research group examining the effects of PCBs and other chemical contaminants on human health.

Mindy Thompson Fullilove, M.D., is a research psychiatrist at the New York State Psychiatric Institute and associate professor of clinical psychiatry and public health at Columbia University. She serves on the editorial boards of several journals, including the *American Journal of Psychiatry* and the *Journal of Sex Research*. Her most recent book is *The House of Joshua: Meditations on Family and Place.*

John Wooding   is associate professor and chair of the Department of Regional, Economic and Social Development at the University of Massachusetts, Lowell. He is a political scientist whose major interests and expertise include the politics of occupational health and safety and environmental regulation, international political economy, and the politics of developed countries. Dr. Wooding is co-editor of a book on occupational health and safety, *Old Problems, New Solutions*.

Index

Porter/Novelli firm survey, 159–160

Positive Outreach Leaders (New Orleans), 308–309

Poverty, 2, 88–89, 232, 235, 249, 276–277. *See also* Racial injustice children in, 3

Precautionary principle, 32–33, 37–38n.41

President's Council on Sustainable Development (PCSD), 182, 197n.12

Procter & Gamble, 162, 188, 206

Proctor, Robert, 28

Production, 46, 150, 317. *See also* Industrial ecology; Social relations of production; *and by industry*

Propaganda and democratic society, 157–160, 171–172. *See also* Public relations industry

Property rights and zoning, 139, 183, 261
corporations challenging local zoning, 262

Psychology of place. *See* Ecological psychology

Psychosocial effects of toxic contamination, 104, 120. *See also* Ecological psychology

Public Affairs Council, 166

Public education on health, 24–29, 37n.30

Public health. *See* Health and the environment

"Public Interest Pretenders" (*Consumer Reports*), 164

Public Media Center, 185

Public opinion, 159–160, 220–221
mining industry and, 262–263, 272–273n.21

Public relations industry, 157–159, 170–174
co-opting environmental images and practices, 183–186, 189–190, 201–204, 208–209, 226
"environmental" PR consultants, 167

"greenwashing" (environmental ads by polluters), 9, 10–11, 157, 183–186, 188, 201, 204, 212, 228, 263
"integrated communication strategies," 187–188
mass media campaigns, 262–266
organizing "grassroots" front groups, 162–165, 172, 173
publicizing industry sponsored research, 160–161, 162
"third party technique," 159
use of "independent experts," 159–162

Public schools, corporations and, 188–189

Racial injustice, 2, 60, 62, 89, 249, 276–277, 302. *See also* Environmental justice movement; Poverty; Toxic environments
environmental racism, 304, 305, 311nn.1 and 3, 311

Rampton, Sheldon, 157

Randels, Kathy, 308

Reagan, Ronald, Executive Order 12291, 138

Recycling, 147, 207–208, 210, 219–220

Redevelopment of toxic waste sites. *See* Brownfields remediation efforts

Redkin products, 215

Redlining practice, 61, 75n.6

Regulatory decisions, 116–117, 182–183, 245–246, 252. *See also* Government policy

Remediation programs
costs associated with cleanup, 60–61, 65
federal and state policies regarding, 63, 65, 70
substandard toxic waste cleanups, 98–100, 107–108
Superfund cleanup sites, 29, 58, 61–62, 97–100, 258

Repetitive strain injuries (RSI), 49

Reproductive labor, 222